FROM NEWTON'S LAWS TO EINSTEIN'S THEORY OF RELATIVITY

FROM NEWTON'S LAWS TO EINSTEIN'S THEORY OF RELATIVITY

Fang Lizhi
Chu Yaoquan

SCIENCE PRESS
Beijing, China

WORLD SCIENTIFIC
Singapore

1987

Published by

Science Press
Beijing, China

World Scientific Publishing Co. Pte. Ltd.
P. O. Box 128, Farrer Road, Singapore 9128

Translator: Huang Hongquan
Responsible Editor: Jiang Shuhua

Copyright © 1987 by Science Press and World Scientific Publishing Co. Pte. Ltd.
All rights reserved. This book or parts thereof, may not be reproduced in any form or by any means, electronic or mechanical, including photocopying, recording or any information storage and retrieval system now known or to be invented, without written permission from the Publishers.

ISBN 9971-978-36-9

Science Press Book No. 4264-96

Printed in Hong Kong

CONTENTS

PREFACE 1

CHAPTER I FROM ARISTOTLE TO NEWTON 3
Concepts of Space and Time —— Aristotle's Centre of the Universe —— The Relativity and Absoluteness in the Newtonian Space-time Concept —— The Critique of Mach

CHAPTER II SPACE, TIME AND MOTION 9
The Measurement of Time —— The Measurement of Length —— Events and World-lines —— Relativity of Motion —— Composition of Velocities

CHAPTER III FROM THE CLASSICAL COMPOSITION OF VELOCITIES TO THE CONSTANT SPEED OF LIGHT 17
The Puzzling Light Phenomena —— Supernova Explosion and Light Velocity —— The Hypothesis of the Ether —— The Michelson-Morley Experiment —— The Speed of Light is Constant —— New Law of Velocity Composition —— The Velocity of Light is a limit —— On Superluminal Speed —— The Measurement of c

CHAPTER IV FROM GALILEO'S PRINCIPLE OF RELATIVITY TO THE SPECIAL THEORY OF RELATIVITY 28
Salviati's ship —— Two principles of the Special Theory of Relativity —— Simultaneity is Relative —— Who Shot First? —— Cause and Effect

CHAPTER V THE RELATIVITY AND ABSOLUTENESS OF A ROD AND A CLOCK 36
Space and Time in the Newtonian Space-time Concept —— The Moving Clock's Time Slows Down —— The Lifetime of a Muon —— The Twin Paradox —— The Moving Rod Contracts —— Mr Tompkins' Mistake —— Lorentz Transformation

CHAPTER VI DYNAMIC PROBLEMS **47**
Aristotelean Dynamics —— The Moving Object Moves Forever —— Newton's Mechanical Laws —— Contradiction between Newtonian Mechanics and the Limiting Velocity of Light —— Inertial Mass Changes With Velocity —— Inertia = Energia — a Foundation Stone of the New Epoch

CHAPTER VII FROM THE LEANING TOWER OF PISA TO THE GENERAL THEORY OF RELATIVITY **54**
The Experiment Atop the Leaning Tower of Pisa —— Gravitational Force —— Gravitational Force: The Precession of Mercury's Perihelion —— Universality of the Ratio: Gravitational Mass/Inertia Mass —— The Nature of Gravitation is the "Null" of Gravitation —— Local Inertial System —— What is Gravitation? —— Einstein's Gravitational Field Equations

CHAPTER VIII FROM NEWTON TO POST-NEWTON **64**
Post-Newtonian Corrections —— The Precession of Planetary Perihelia —— The Precession of the Axis of Rotation —— The Gravitational Redshift —— The Deflection of Light —— The Time Delay of Radar Echo

CHAPTER IX FROM CLASSICAL GRAVITATIONAL COLLAPSE TO THE BLACK HOLE **73**
Something More About the Strong Field Requirement —— Gravitational Collapse —— Where are the Strong-field Objects? —— Pulsar is a Kind of Compact Object —— The Structure of Neutron Stars —— The Black Hole —— The Black Hole is Hairless —— Critical Mass —— X-ray of Binaries

CHAPTER X THE CONFIRMATION OF GRAVITATIONAL WAVES **85**
Einstein's Prediction —— Gravitational-Wave Source in the Universe —— Weber's Experiment —— Gravitational Radiation Damping of Binaries —— PSR1913 + 16 —— An Ideal Relativistic Celestial Laboratory —— Gravitational Radiation Damping Tested

CHAPTER XI FROM NEWTON'S UNIVERSE TO THE EXPANDING UNIVERSE **93**
From the Finite Bounded to the Infinite Unbounded —— The Difficulty of the Newtonian Infinite Universe —— An "Idiotic" Problem —— The Finite Unbounded Universe —— The Expanding Universe —— The Big-Bang Cosmology —— The Age of Objects —— Microwave Background Radiation —— Abundance of Helium

CHAPTER XII AFTER EINSTEIN **106**
Seeking After Unification —— The Great Unity and the Very Early Universe —— Gravitation and Quantum Theory —— Black Hole Emission —— Superunification and Singularity

PREFACE

More and more historical facts have proved that the civilization mankind boasts of today is in large measure indebted to the toil of a number of outstanding physicists, of whom Galileo and Newton, pioneers of classical mechanics, and Einstein, founder of the special and general theories of relativity, are representative.

When we recall these epoch-making scientists, we discover, that although each of them lived in an entirely different age, and that in the evolution from classical mechanics to modern physics many great changes took place, yet these scientists all possessed the same traits of character that distinguished them from the rest of us.

First of all, they were all sincere explorers of truth who refused to be led by prejudices or be enthralled by traditional ideas. Their scientific works often commenced with a breakthrough in the long-since-settled doubt-defying postulates. Their vocation, even in the eyes of the modern age, seemed to be "too" fundamental and "too" abstract. For instance, they ventured to ask, "What is time? What is space? What is relativity? What is absoluteness? What are the laws governing the motion of celestial objects? Where did the universe originate from? ..." These problems appear so alien to the daily affairs with which most of us are concerned that they seem to belong to a transcendental world. Yet, as facts have eloquently shown, the development of these "scholastic" debates have already brought about such technical progress as can hardly be replaced by any other.

Secondly, they were all adherents to a method peculiar to the researches in natural science. Though they were most interested in abstract ideas, they never gave free rein to vain sophistry. On the contrary, they always confronted theory with physical experiments or astronomical observations, so as to check each concept and hypothesis contained therein, before deciding whether the theory was sound or unsound.

Thirdly, most of them were not mere scientists in the narrow sense of the word: they were men whose souls were brimming over with devout sentiment and lofty aspiration. For instance, Einstein's life represents a wholehearted devotion to reason, science, and democracy. Once he said: "Only when a man has dedicated himself to society, can he really find the meaning of life, though fleeting and precarious as it is." Resenting submission to tyranny and fear of the scientific spirit, he once asked: "Tell me, if Giordano Bruno, Spinoza, Voltaire and Humboldt all think like this, all act like this, where would we be now?"

If only because of this, both their persons and their ideas could not be tolerated by the rulers of their times. Galileo was persecuted by the Inquisition. Einstein was oppressed by the German Nazi regime. Yet what a ludicrous thing it is, that in the seventies, when those semi-illiterate "gang of four" went all out imposing fascist domination over culture, even Einstein's theory of relativity could not have been spared this merciless persecution! Although the old days have gone by like a nightmare, yet to recall them as we do now is not only helpful to, but also necessary for, the proper comprehension of the actual course of the development of physics. Because, if we are only at home with those academic details of Science, yet cannot grasp her spirit, her very soul that makes her what she is, how can we have transplanted her into our soil and, by cultivating her, let her take root, flower, and bear fruit?

The aim of this little book is to introduce the reader to the main thread of development leading from Newton's laws to Einstein's theory of relativity. Limited by its scope and avoiding as much as possible the use of mathematical apparatus, we have only wished to make clear the most fundamental ideas and concepts. This wish of ours, however, might not have been substantiated or fully substantiated. Here we sincerely hope the reader will point out where and how this little work should be improved.

In the autumn of 1978, urged on by the Editorial Committee of the *Essential Physical Knowledge Series*, we began to plan this little book. Mr Chu Yaoquan and I worked on the first manuscript in stages. In the course of a lecture tour of Italy, in March 1979, I found some time to rewrite the book from the very beginning. During those days, as I lived in the Accademia dei Lincei of Rome, I often lifted my eyes and saw through the window the emblem of the academy, the same emblem printed on the cover of Galileo's *Dialogue*. Though the emblem of the most ancient academy looks so time-worn, yet whenever I write something about Galileo, it cannot but evoke in me a deep reverence for him. I remember that the day following the completion of the manuscript was March 14, 1979, the one-hundredth anniversary of Einstein's birth. The Chinese edition of this book was published by Science Press in April 1981. Since then, it has been reprinted twice. This English edition is translated from Chinese, but some additions have been made in the final part. I offer it as a bequest to the readers. Let me humbly dedicate it to Galileo, Newton and Einstein, the giant stars of science, who are worthy of our esteem and admiration for ever and ever....

Hefei
Autumn 1984

Fang Lizhi

CHAPTER I   FROM ARISTOTLE TO NEWTON

Concepts of Space and Time

Some physical concepts appear to be rather plain, though they are not necessarily so simple.

For instance, "At eight o'clock this morning I started to read at home." Such a sentence is often heard in daily speech. Yet even in this common sentence, two basic concepts are involved. "At eight o'clock this morning" denotes the time, while "at home" refers to the place, or the position in space. Space and time are the most plain and most frequently used concepts. Yet what is time?—if one asks—and what is space? To these two questions we shall find it not easy to give adequate answers. Yes, these two questions as mentioned above would defy any ready answers. Though people have tried, in one way or another, to define space and time, none of those definitions have proved to be sufficiently satisfactory.

In attacking a physical problem, the correct approach is often to start not with the precise formulation of definitions for the concepts relevant to the problem, but with the analysis of the true relations between the said concepts. Therefore, as far as the basic concepts of space and time are concerned, the most essential thing is not to give "perfect" definitions to them, but is the demonstration of the innate relationships between space, time and the motion of matter.

By way of illustration, let us again examine the above sentence. When we say "eight o'clock in the morning," we may be referring to the watch we are wearing or the clock we use at home. To be more exact, we are using Peking standard time as the reference. Obviously this specification of time is only relative. If we use Tokyo standard time, it will be nine instead of eight o'clock. This is a kind of relativity in specifying time--a certain attribute of time.

Now if another man says: "Hey! I also started to read at eight o'clock this morning," we shall readily come to this conclusion: these two men started to read at the same time, i.e. eight o'clock of Peking time. In this case "common sense" leads us to believe that the "simultaneity" of these two events is absolute instead of relative. That is to say, if two events, according to one clock, take place at the same time, their simultaneity is considered a matter of course according to all other clocks. This habitual judgement, however, is not always right. Though it may be right to say that these two men started to read at the same time according to Peking or Tokyo standard time, an observer in a high-speed spaceship may find by his watch that they did not actually start to read at the "same time." Simultaneity is not absolute, but relative. To one observer they may occur at the same time, yet to another, they may not. This assertion which is "contrary" to habitual conception constitutes the essential difference between Einstein's concept of space and time and

Galileo and Newton's classical concepts of space and time.

As to the concept of space and time, it is the understanding of their physical property we have attained. Any great change effected in the history of science is often accompanied with the birth of a new concept of space and time; and the vice-versa, in a certain sense, is also true: the change in the space-time concept is one of the most essential features that mark the great revolution of science. Therefore if we want to understand the spirit of the successive historical stages of the development of human knowledge, from Aristotle to Newton, and then from Newton to Einstein, we must, first of all, understand their respective concepts concerning space and time.

Aristotle's Centre of Universe

Aristotle's concepts of space and time had dominated over the European mind since his time up to the Middle Ages. The Aristotelian theory insists that the earth is the centre of the universe. The whole universe consists of seven hollow spheres of different sizes with the earth as their common centre; and the moon, the sun, the planets and the stars are studded, in succession, on different spherical shells, and they all move in a perfect circle. Today few would accept a cosmological model like that. Of course, equipped with the scientific knowledge that modern age boasts of, we may find it quite easy to criticize Aristotle. But after all yesterday is not today. Just imagine that more than two thousand years ago, Aristotle had ventured a unified explanation of the universe, asserting that the earth looks like a ball! This alone, we

Fig. 1–1 The earth conceived before Aristotle

have to admit, is a great step forward in the development of human knowledge, since in the ancient mind the earth had been conceived as a flat thing placed on the back of a huge turtle that floats on the sea. Then how could it conceive the earth as a globe? According to the "conventional" idea of that time, those people who inhabited the regions of the globe diametrically opposite to ours should have "fallen" into the bottomless space long ago. Reasoning like this, we can imagine what mental reluctance, born of prejudice, such a global notion of the earth had to overcome before it was finally established! As far as the notions of space and time are concerned, Aristotle had relativized "up" and "down". When we think the people on the opposite hemisphere living "underneath" us, they also flatter themselves with the same fancied superiority. This subjective thinking only points to the isotropy of space, i.e., among all the possible directions none can "get the upper hand" of the rest. This notion, understood as the relativity of direction in space, is an important step mankind takes towards the scientific comprehension of space and time.

In the Aristotelian system, the position an object occupies in space is of paramount importance. Positions in space are absolute, and the centre of the earth is believed to be the centre of the universe. Every object has its own natural position. If no obstacle is placed in its way, it will manage to get to the position allotted to it by nature. The reason why an object is in motion is only because it has not yet attained to its natural position. Aristotle divides the space of the universe into two parts: that "above the moon" (farther away from us than the moon), and that "beneath the moon" (nearer to us than the moon). The natural positions of the celestial objects such as the sun, the moon and the stars are fixed on different layers of the celestial globe, and as the latter moves in a circle, they move likewise with it. The natural position of all the objects near the surface of the earth is the centre of the earth, and that accounts for their tendency to fall downward to the ground. As Aristotle conceives of space and time, some position (such as the centre of the earth) boasts of a peculiar character. In the realm of all natural laws governing the motion of objects, particular points in space as such are believed to play decisive roles. This characteristic of space may as well be called the absoluteness of the spatial points. In modern language we shall say that the Aristotelian space maintains that all directions in space are equivalent but inhomogeneous, since different positions in space play different roles.

The Relativity and Absoluteness in The Newtonian Space-time Concept

Aristotle's concept of space and time was evolved on the basis of the knowledge of nature the ancient Greeks had accumulated: it fell in with the explanations people then gave to natural phenomena. As the old knowledge gradually gave way to the new in the course of the development of science, the old concept of space and time evolved accordingly into to a new one.

The new science, of which Copernicus, Galileo and Newton were the founders rejected what characterized the Aristotelean space-time concept-- the absolute spatial positions in his system. Copernicus refuted the absolute significance of the earth viewed as the centre of the entire universe. Galileo unequivocally proclaimed the relativity principle (which will be discussed in Chapter 4). Newton even tore down the partition between the heaven "above the moon" and that "below the moon", believing that the falling of the apple and the revolving of the moon round the earth were both effected by the same cause—some other cause than their desire to return to their "natural position." Therefore in Newton's equations no point has the privilige of being the centre of the universe: all space-time points are equal. Physical laws remain the same no matter which of the space-time points is to be investigated. This is the relativity in the new concepts of space and time.

Yet, one can only walk step by step along the way to success. Though Newton stood looking over the shoulder of Aristotle, his mechanics still bases on the concepts of the absolute rest of space and the absolute changelessness of time. In his *Mathematical Principles of Natural Philosophy* Newton wrote: "Absolute space, insofar as its character is concerned, has nothing to do with any of the exterier conditions: it forever maintains regularity and sameliness." "Absolute, pure and mathematical time, insofar as itself and its character are concerned, flows evenly, unaffected by any exterier conditions." In short, in Newton's space-time concept, space, time, and the "exterier conditions" are mutually independent, since spatial extension and temporal flow are absolute. In this sense we may say that Newton's system has preserved the Aristotelean absoluteness.

To visualize "space" as a stage on which objects make their mechanical movements, or to visualize it as the background of their actions, is a very natural outcome of intuitive abstraction. In our daily life we never wonder at the experience that in a certain suitcase we can store a certain amount of goods. This propety of the suitcase may be termed as its volume or space. The measurement of the volume or space of the suitcase has nothing to do with what objects are contained therein (it remains the same even when the suitcase is empty). As a rule the seller of the suitcase never forgets to mark its size, such as 26×20×10. That he can always succeed in marking it is based on the fact that the measurement of its volume is never affected by "exterier conditions." Now if we let the suitcase expand infinitely, what we get will be a typical Newtonean absolute space--a space independent of any particular substance contained in it.

Isaac Newton was an empiricist. He could not allow any *a priori* notion to enter into his system. To him the physical entities must be those which can be grasped sensually. Yet how can our senses grasp the space which he defined as "absolute"? Here Isaac Newton devised an ideal experiment—the famous rotating "bucket" experiment by which what motions are absolute with reference to the absolute space can be decided. A bucket filled with water is made to rotate. At the beginning the bucket rotates without dragging the water along into a likewise motion. Though the water moves relative to the inner wall of the bucket, its surface remains level just as when it is still. Gradually the water is dragged into motion by the bucket wall and begins to rotate along with the latter. Though no relative motion exists between the water and the bucket wall, a concave appear in the water surface: the water near the rim is higher than that in the centre. Therefore even when no relative motion takes place between the water and the bucket, we may just as well decide whether the bucket as a frame is in rotation or not with reference to the absolute space: if the water surface remains level, the bucket is not in absolute motion; if the surface becomes concave, it is in absolute motion. This is the touchstone Isaac Newton offered us to decide the case.

Up to now the Newtonian system appears to be faultless, though things may not be what they seem.

The Critique of Mach

The concept of absolute space brought forward by Newton has caused, ever since its inception, many scientists and philosophers to contemplate and to question its validity.

If there really existed such an absolute space as differs from all other spaces, then the motion of each and every object relative to this particular space should be measurable, i.e., every object in its absolute motion must have a definite measurable absolute velocity. On the other hand, if none of the physical laws contains such an absolute velocity, then we shall find it impossible to measure it. It we cannot measure it, we shall fail to experience its absoluteness. Such laws wherein the absolute velocity is involved will treat the direction parallel to the absolute velocity differently from directions perpendicular to it. In other words, the space may have different characters in different directions, ie the space is anisotropic. Not only the conception alone will cause theoretical confusion, but experiment has never revealed such a thing as anisotropy of space.

Leibnitz, then Berkeley and, then Mach, have all criticized the concept of absolute space-time. Their analytical critique, mainly based on philosophical argument, blazed a new trail for the progress of the space-time concept. That philosophical speculation can contribute a great deal to the scientific progress is most typically exemplified in the works of Mach and others. Many of the eminent physicists set great store by the philosophical speculation, which fact is most obvious in the history of the development of the space-time concept.

Mach in his *Die Mechanik*[1] wrote: "If we say an object K can only change its direction and speed after being acted upon by another object K', then, when other objects A, B, C⋯, by which we can decide the motion of the object K, do not exist, we can by no means obtain such an impression. Therefore, what we actually know is only the relations between the object and A, B, C⋯ If now we suddenly wish to neglect A, B,C⋯, and talk about the behavior of the object K in the absolute Space, we'll then commit a double mistake. In the first place, in case A, B, C⋯ do not exist, we shall not be able to know how the object K will act; secondly, we shall find ourselves deprived of any means to determine the action of the object K, whereby to testify our assertion. Any assertion as such will lose all its import as natural sci-

---

(1) Title of the English translation is: *The Science of Mechanics, A Critical and Historical Account of its Development* (Open Corut, La Salle, I11., 1960).

ence." In this manner Mach argued for the impossibility of describing a motion with reference to the so-called absolute space devoid of matters (A, B, C···). Absolute space is devoid of scientific significance.

In accordance with this viewpoint, Mach insisted that the concave water surface in Newton's bucket does not betray the rotation of the bucket with reference to the absolute space, but reflects the rotation of the bucket relative to the earth and other celestial objects. The concavity arises not from the absolute rotation of the bucket but from the existence of various objects in the universe that act upon the bucket. Whether the bucket rotates relative to the other objects in the universe, or the other objects rotate relative to the bucket, the result will be just the same: the water surface will become concave. Therefore, the concave water surface can only prove the relative motion between the bucket and the other objects in the universe (A, B, C···). It does not point to the existence of the absolute space. Mach's analysis of the bucket experiment shows that he not only took uniform motions as relative (since no absolute velocities exist with respect to the absolute space), but also regarded accelerated motions as relative (since no absolute accelerations exist with reference to the absolute space). This latter idea exerted a great influence on the formation of Einstein's general theory of relativity.

Commenting on Mach, Einstein once said: "Mach's historical critique has an enormous influence on the scientists of our generation." For that idea of Mach that inertia and inertial force originate from the interaction between the material entitities of the universe, Einstein and others have coined a term: Mach's principle. Though Mach, even in his old age, never thought himself as the forerunner of relativity, yet judging from his critical analysis of Newton's concept of space and time, we can say objectively that his enlightening idea really helped the birth of the relativistic concept of space and time.

From Aristotle's relativity of spatial direction, to the relativity of positions and moment in Newtons's system, and then to Mach's critique of absolute space-time, scientists of successive generations have gradually freed themselves from the human prejudices concerning the nature of space and time, and from the falsehood of the a priori absoluteness that has hitherto passed for truth. Of this general trend of history Einstein's work is a brilliant continuation. In Einstein's relativistic concept of space and time, those eroneous ideas to which our common sense is accustomed are more resolutely discarded. The history of the development of the space-time concept makes us fully aware of the fact that since we all live in a very limited space-time realm, we even cannot properly understand space and time in which we live. This is the first good lesson the study of space-time physics can teach us.

In the following chapters we shall discuss the details of the historical transition from Newton's classic mechanics to Einstein's theory of relativity. While reading these chapters, please now and then recall the aforesaid lesson: trust science and not those ideas that seem to have occupied your mind since infancy.

CHAPTER II   SPACE, TIME AND MOTION

The Measurement of Time

This chapter will mainly aim to provide for some preparatory knowledge and will begin with the simplest things.

Physics is an experimental science, its laws sum up experiments, especially the quantitative experiments. Therefore, in studying the physical problems of space and time, we should first of all understand how space and time are measured.

When we talk about the measure of time, the subject will naturally remind us of the clock or the watch, though they are not the only devices by which time is measured.

In 1583 a youth from Tuscany happened to take interest in the swinging motion of a chandelier hanging from the ceiling of a grand church hall in Pisa. He decided to make a study of the laws governing the pendular motion. At that time the first clock had not yet come into use, not to say a stop-watch. As the chandelier swings rather fast, by what means can the short period of the swing be measured? The young experimentist devised a method: he pressed a finger of one hand on the wrist of the other where the pulse could be felt. He watched the motion of the chandelier and mean while counted his pulse. Then he discovered a law: the amplitude may vary, yet during the interval the pendulum completed a to-and-fro swing the number of pulses remained the same. In other words the period of a pendulum has nothing to do with the amplitude of the swing. This famous observation may be called the first experiment of scientific physics. This intelligent experimentist who laid the foundation of physics is—Galileo.

Galileo's method tells us what is the crucial thing in measuring time. In principle any periodic process can be regarded as a time-measuring instrument--a clock. In nature such periodic processes abound, and some of them have long since been employed by our fathers as the time-measuring standard. For instance the rise and fall of the sun marks a day, the cyclic motion of the four seasons indicates a year, the wax and wane of the moon records a lunar month. All these are familiar things to everyone of us. Other cyclic processes such as the revolution of the double star, the beating of the human pulse, the swinging of the chandelier, the vibration of the molecule, etc, can all be used as the time-measuring standard. In short, thousand and thousand of types of periodic motion in the world can serve as the "clock." Of course, a clock may be good or bad. Comparing the pulses of two men, you will often find an obvious difference between them. Therefore pulse cannot serve as a good clock: it is not so stable. If you compare the periods of two simple pendulums, you will find that they are more stable. The stability of the pulsar's pulsating period is much more satisfactory. Before 1967, the rotation of the earth had been considered as the best time-measuring standard. After 1967, a more stable "clock" has been

Fig. 2–1 Galileo measures the period of the swinging chandelier

adopted as the standard, i.e. the period T of the microwave radiation between the ground-state hyperfine structures of ^{133}Cs has been taken as the unit of time, the relation between T and the second being

$$1 \text{ s} = 9\ 192\ 631\ 700 \text{ T}$$

The Measurement of Length

The basic length-measuring instrument is the ruler. In measuring a straight line between two points, the method is to start measuring foot by foot, from one point till the other is reached. In the Tang Dynasty the astronomer Zhang Sui and his collaboraters measured with a stretched rope, section by section, the distances between Kaifeng, Huaxian and Shangcai in Henan Province, his aim being to determine the length of one degree along the meridian. In human history this is probably the first measurement of length on a large scale in strict accordance to the primitive way of measuring length with a ruler.

As for rulers, there are various kinds. Anything of a definite length can serve that purpose. A certain part of the human body can be used as the length unit. The word *foot* in English is a length unit, since for a time the human foot served as a unit of the linear measure. Just as in the case of measuring time, a good ruler has to be chosen as the common standard in measuring length. Since rulers made of any material are more or less subject to the influence of the environment, they cannot meet the requirement of a standard. For this reason, the model bar of metre in Paris that has hitherto been honoured as the international standard is no longer in use. To replace it the process of light emission of atom has been employed recently, i.e. the wave length λ of light emitted during $^2P_{10}-^5d_5$ transmission of ^{86}Kr is used as the standard unit[1]. The relation between the metre and λ is

$$1 \text{ m} = 1\ 650\ 763.73\ \lambda$$

In discussing the problems of the solar system, we can take the mean distance between the earth and the sun as the unit, and the length of that "ruler" is called 1 astronomical unit (A.U.). For the distance between stars, the light-year is often used as the unit. One light-year is the distance that light travels in one year, i.e. 9.5×10^{15} m approximately. For instance, our nearest neighbour star is about 4 light-years away from us. The figure not only tells us how great the distance is, it also tells us that the star we see today from the earth is what it was four years ago, for the light emitted there from 4 years ago reaches the earth just now. That is to say, when we are scanning the starry sky, the farther our sight penetrates, the earlier things we see. This fact already gives us a hint that time and space are often related.

---

(1) More recently, light speed c has been taken as a stendard to replace the stenderd nnit of length.

Events and World-lines

Now that we have defined the measurement of time and length, we may begin to study the motion of objects.

As for motion, as far as a macroscopic object is concerned, it is a sequence of events marked out by space-time points. In a railway time-table, you can find a series of names of stations at which the train will arrive in succession. The name of a station and the time the train arrives at it constitute an event.

The motion of the train consists of the aforesaid events. Generally speaking, the time and the place jointly constitute the event. The motion of a macroscopic object can be depicted as a sequence of events, and the latter are the essential constituents of the former.

station name	distance measured from Kunming (km)	Kunming-Shanghai express
Liuzhou	1,246	08 20.04
Yishang	1,157	26 18.16
Jinchengjiang	1,085	17.00 16.45
Nandan	984	17 14.12

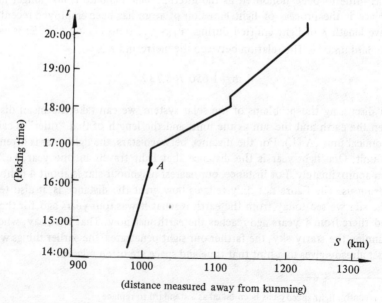

Fig. 2-2 The motion of an object represented by a world-line in the space-time diagram

We may also use a diagram to describe the motion of a train. In the following diagram, the horizontal coordinate denotes how far away the travelling train is from Kunming, the perpendicular axis denotes the Peking standard time. The diagram is therefore called a space-time diagram. Each event, i.e. a certain place in association with a certain time, appears as a point on the diagram. For example, point A denotes the event that the train is at Jinchengjiang at 16.45. In the diagram the motion of the train is represented by a line. When the train stops at a station, a straight line parallel to the time-axis is used to denote the event, since during the train's stay at the station, the horizontal coordinate of the train (its position) remains unchanged while time is flowing past as usual. The line in the space-time diagram is called the world-line. Every motion has its corresponding world-line on the space-time diagram.

Relativity of Motion

In the above-mentioned train schedule, the distance is measured with Kunming

Fig. 2–3 The relativity of the mode of motion

as the starting point and the time is Peking time. If "measured-from-Guiyang" distance is adopted, or some other standard time is used instead, then the figures in the schedule will be entirely different. That is to say, if a different method of denoting space and time is used, the data of the event will be different. Relativity of this nature has already been mentioned in the previous chapter.

While the description of an event is relative, the very mode of motion is not absolute either. That is, to different observers, the same motion can assume different appearance.

Suppose that on a rainy yet windless day, two observers, K and K', are studying the tracks of the falling raindrops. K stands motionless. In his eyes the raindrops are falling perpendicularly. For this reason he holds his umbrella erect. K' is walking at a quick pace. In his eyes the raindrops are falling aslant towards him. Therefore he always holds his umbrella slantingly forward to prevent himself from getting wet. Therefore, when somebody wants you to describe in what direction the raindrops are moving, let him tell you first, to which observer the question refers. With reference to no particular observer, the question itself does not make much sense.

Usually we call an observer, i.e. a certain definite measurement of space and time, the frame of reference. The conclusion of the above analysis is: relative to frame K, the raindrops are falling straight downward; relative to frame K′, the raindrops are moving slantingly. This is the relativity of the mode of motion.

Composition of Velocities

Velocity is a physical quantity that indicates both the rate of motion of an object and the direction in which the object moves. Velocity is also relative. With respect to different frames of reference, i.e. different observers, the velocity of the same object may be different. Therefore we can only discuss the speed of a moving object and the direction in which it moves with respect to a definite frame of reference, otherwise the discussion is meaningless.

Now let us turn again to our K and K'. When K' walks with ever quickening steps, he not only sees the slanting rate of the raindrops growing (their direction changing), but also feels their velocity increasing. Whoever has driven in a convertible may have the same experience as that of the observer K'.

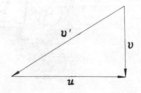

Fig. 2–4 Composition of velocities

In the following passages we shall describe quantitatively the relation between the velocity of the raindrops and that of the observer. In the following diagram, the downward perpendicular arrow v denotes the velocity of the raindrops as observed by K (the direction and the length of the arrow represent respectively the direction and the magnitude of the velocity). The horizontal arrow u represents the velocity of the observer K' relative to K. The arrow v' indicates the velocity of the raindrops relative to K'. u, v and v' constitute a triangle. Therefore, as long as the velocity of K' relative to K is increasing, the velocity of the raindrops relative to K' is also increasing. Expressed mathematically, the statement is

$$v = v' + u$$

The rule shows that the velocity of the raindrops is related with the motion of the observer--an aspect of the relativity of velocity.

The relativity of velocity has another aspect. At the athletic game, the javelin thrower always runs some distance before he thrusts out the missile. This is because, if the velocity of the javelin on its initial flight relative to the athlete (frame K') is v', the velocity of the running athlete relative to the ground (frame K) is u, then the velocity of the javelin relative to the ground is $v = v' + u$. Therefore the athlete's run helps to increase the velocity of the javelin relative to the ground. In other words, the velocity of the javelin relative to the ground is related with the thrower's

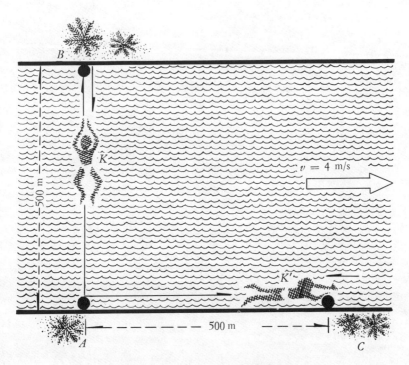

Fig. 2–5 Swimming contest in a little river

state of motion. This is another physical law.

These two aspects of the relativity of velocity, if put together, will be termed as the composition of velocities.

The composition of velocities would appear as a matter of course to many people. Actually in daily life we have experienced the thing thousands and thousands of times and have made use of it very frequently. When you swim in a rapid current, no matter how hard you try to arrive at the directly opposite bank, you will only land somewhere down the river. In this case the law of velocity-composition has secretly played its role.

To test the reader's understanding of the formula of velocity composition, let him try to solve the following problem. A little stream 500 metres wide, flowing at the speed of 4 m/min. Two swimmers K and K', who swim at equal speed in still water, i.e. 50 m/min, both start from point A of the bank. K swims to the directly opposite place—point B and then returns to the starting point. K' swims down-stream from A to C (the distance in between is also 500 metres) and then returns to the starting point. If these two swimmers leave point A at the same time, who will return first, and how much earlier?

Here the reader may get somewhat bored, since the problem seems to be too plain and obvious to need a lengthy explanation. But, as marks a certain feature of physics, no "simple" conception whatever can be let pass unexamined. After carefully examining it, we shall probably find that the "plain" fact is not really so plain.

CHAPTER III FROM THE CLASSICAL COMPOSITION OF VELOCITIES TO THE CONSTANT SPEED OF LIGHT

The Puzzling Light Phenomena

The previous discussion centres upon one of the most important rules of the classical mechanics— the composition of velocities. In this chapter we shall discuss how the conception of velocity composition develops along with the transition of classical mechanics into the theory of relativity.

First of all let us point out that the classical law of velocity composition is correct only within certain limits.

Now let us analyze, with the help of the velocity-composition formula, some phenomena in connection with the propagation of light. According to optics, the reason why we can see an object is because the light emitted (or reflected) by the object enters our eyes. An object that neither emits nor reflects nor absorbs light is invisible.

The following picture shows K and K' at the play of passing a ball. Now K throws and K' receives. The reason K' sees the ball is because the light emitted (or reflected) by the ball reaches K'. Let c denote the speed of light and d the distance between K and K', then the instant K' sees the ball at rest in K's hands immediately before he throws it will be $\Delta t = d/c$ later than the instant the throwing action actually begins.

Now let us say that the initial velocity of the ball is u. If the motion of light also satisfies the classical law of velocity composition, just as the motion of the athlete's javelin does, then the velocity of light emitted by the moving ball should be a little greater, i.e. $c+u$. Then the time K' sees K throw the ball should be $\Delta t' = d/(c+u)$ later than the ball actually leaves his hands.

Compare Δt with $\Delta t'$ and it will be discovered that since $c+u>c$, $\Delta t'<\Delta t$. That is to say, K' will first see the ball leave his partner's hands and then see it stay in the latter's hands. More dramatically speaking, K' will first see the ball on its flight, then see the throwing action! Such is the confusion ensuing from the application of the formula of velocity composition to the problem of light propagation we first see the event that happened next and then see the event that happened first. Yet such topsyturvy of time sequence has never been witnessed by anybody. This proves that light does not fit in with the velocity-composition formula!

Here some may say that the velocity of light is so great, that Δt and $\Delta t'$ are pratically not much greater than zero. Therefore even if $\Delta t'<\Delta t$, the disparity cannot be noticed. Indeed the velocity of ordinary objects witnessed in daily life is usually very small compared with that of light. Therefore if the velocity of light is regarded as an infinity, the aforementioned contradiction will disappear. Yet on astronomical scale it cannot be treated as an infinity, and the aforesaid contradiction in the

Fig. 3-1 Strange phenomena in the ball play

propagation of light cannot be avoided. The following is a real case.

Supernova Explosion and Light Velocity

More than 900 years ago, a well-known supernova explosion was recorded in detail by Northern-Song astronomers. According to a historical record, the explosion began in the 5th month of the first year of the Zhihe Reign during Renzong's time, i.e. 1054. In the first 23 days the star was so bright in the sky that it could be seen even in daytime. Then it gradually grew dim until in the third month of the first year of the Jiayou Reign (1056) it became invisible to the naked eye. The explosion lasted 22 months, and its remnant is the famous nebula located amidst the constellation of Taurus, the Crab Nebula as it is called.

This ancient record can be connected with the investigation on the speed of light. When a supernova explodes, the outer shell of the star will fly outward in every direction. Some fragments will move towards us (position A in Fig. 3-2), while some will move in a perpendicular direction (position B). If light obeys the above-said law of velocity composition, then by the same analysis as has been made of the ball play, we shall conclude that the velocity of light emitted at position A is $c+u$ while the velocity of light emitted at position B remains close to c.

Therefore, the time taken by light from A to reach the earth should be $t = L/(c+u)$, while the time for light from B is $t \approx L/c$. The distance between the Crab

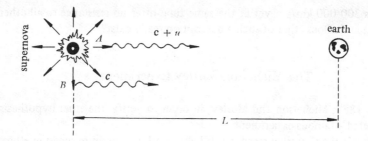

Fig. 3-2 Light propagation of an exploding supernova

Nebula and the earth is about 5 000 light-years; the speed of the flying fragments is about 1 500km/s. With these data, the result can be readily obtained

$$t' - t \approx 25 \text{ years}$$

That is to say, within about 25 years we should have been able to see the strong light emitted during the initial stage of the explosion. Yet this does not agree with the fact. According to the historical record, "It dims in just over a year." The fact disproves the computation. The conclusion is that the velocity of light emitted either at A or at B is the same. In other words, the velocity of light has nothing to do with the velocity of the object that emits it: no matter how fast the light source travels towards us, the velocity of light remains unchanged with respect to us. Light does not obey the classical law of velocity composition.

The Hypothesis of the Ether

For the phenomenon discussed above people thought of another explanation.

If we make a careful study of the ship sailing on the sea, we shall discover: the speed of the waves produced by the ship generally does not depend on the speed of the ship. It is because, with respect to a certain sea condition, the speed of wave propagation remains constant, unaffected by the speed of the ship.

Here, by analogy, we naturally think that light might be waves moving in a certain kind of "sea". Its velocity is solely decided by the character of the "sea", independent of the speed of the light origin. Truly, light possesses wave properties, and this contributes to strengthen the "sea" theory. Therefore in history the theory had enjoyed wide popularity for a time. The "sea" in which light travels is usually called ether. Since light can reach everywhere, the whole universe is supposed to be permeated with ether. This ether born of imagination, except that it serves as the medium for light propagation, is invisible to the eye, and cannot be experienced in any other manner. In order that it can explain the various attributes of light propagation, ether is required to possess a number of special properties. For instance, it has to be exceedingly rigid, so as to allow light to travel in its medium at a speed as high as 300 000 km/s , yet at the same time offer no resistance to all other objects moving in it. Does ether of such a description really exist?

The Michelson-Morley Experiment

In 1887, Michelson and Morley, in order to verify the ether hypothesis, jointly conducted a famous experiment.

This is the way they reasoned: if light travels at a definite speed in ether, then to a receiver who moves at a certain speed relative to ether, the speed of light coming to him in different directions should be different accordingly. The speed of light that shines on his face should be greater than the speed of light that shines on his back. The difference of these two speeds, if it can be measured, will give support to the ether hypothesis.

Compared with the enormous speed of light, the speed of ordinary objects is invariably very small. Therefore, even if the speed of light differs in different directions relative to an object, it can hardly be measured. The ingenuity of the Michelson-Morley experiment lies in that, instead of measuring the actual values of the light speed in different directions, it aims at measuring the difference of the light speeds in different directions.

The experiment is devised as follows. When a beam of light originating from S reaches the semi-transparent mirror A, part of it penetrates the mirror, while part of

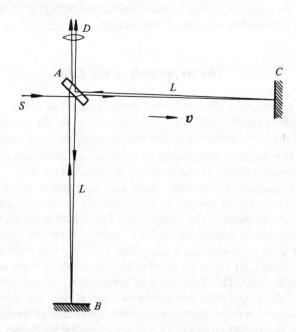

Fig.3–3 A sketch of the Michelson-Morley experiment

it is reflected. The transmitted beam reaches mirror C, reflected, returns to A where part of it is reflected and goes to lens D. In the meantime the other branch reflected by mirror A reaches mirror B and then returns to A where part of it penetrating the mirror goes to D. Now let us assume that the earth is moving in the direction of SC at speed v relative to the ether. Then the journey along $A-C-A-D$ line will take a different time from that along $A-B-A-D$ line. This particular case resembles the exercise given at the end of the previous chapter: light travelling along $A-C-A$ corresponds to swimmer K', light travelling along $A-B-A$ corresponds to swimmer K. An easy computation shows the difference of travelling time of these two beams is

$$\Delta t \approx \frac{L}{c} \frac{v^2}{c^2}$$

where L is the length of $\overline{AC}$ or $\overline{AB}$. By virtue of the interference of two beams of light, the time disparity can be measured.

But the experiment only gave a negative result: any Δt other than zero has never been obtained. Now two alternatives are left to us: either the speed of the earth relative to the ether is zero, or the hypothesis of the ether is wrong. The first solution is unacceptable, since relative to the sun, the earth has its orbital motion besides the rotation about its own axis. Moreover, relative to the centre of the Milky Way Galaxy, the solar system as whole is also in motion. How can we regard the speed of the earth relative to the ether as zero? If we accept this point of view, does it not mean that we regard the earth as a celestial object located at a very special position? Since the time of Copernicus, people have long abandoned any conception that presupposes, no matter in what manner, the superiority of the earth as the centre of the universe. Therefore the only conclusion left to us is: the hypothesis of the ether is wrong.

Thus the conception that light moves in the imaginary ether was irrevocably overthrown. Since this conclusion is so important, many scientists afterwards have repeated the Michelson-Morley experiment in different seasons, at different times and along different lines of measurement. In recent years the employment of laser has greatly improved the precision of the experiment, yet the conclusion remains unaltered.

The Speed of Light is Constant

The important role of the theoretical work is to draw from particular experiments a generally applicable conclusion. Since all particular experiments are completed under particular conditions, they can only be endowed with universality through theoretical abstraction.

In the above analysis the negative experimental result demonstrates that light does not obey the classical law of velocity composition. From this particular result, what universal conclusion can be attained, then? The conclusion is: the speed of light is constant, or the speed of light is characterized by that absoluteness. By the absoluteness of the light speed, we mean when light is travelling in vacuum its speed is constant, unaffected by the motion of its source.

Let it be emphasized again that, so far as a universal law is concerned, in principle we cannot say that the law has been verified by experiment, since a universal law should be applicable to infinite particular cases, yet in our finite span of time, we can only complete a finite number of experiments. Therefore it is less adequate to say that the experiment has testified the constancy of the speed of light than to say that the conclusion drawn from the scientific experiments is in no case contradictory to any of the experimental results we have hitherto obtained.

The constancy of the speed of light makes it greatly different from the speed of any ordinary object. In the previou chapter we have emphatically pointed out the relativity of velocity, i.e. we can only discuss the magnitude of velocity with respect to a certain frame of reference. Yet the speed of light is an exception. As regards a certain beam of light, its velocity remains c, with reference either to the observer K, or to the observer K'.

Just as in many cases we have unwittingly employed the classical law of velocity composition, so have we applied the constancy of the speed of light to some practical cases. The measurement of the distance of an object by radar is an example. If the interval between the emission of the radar signal and the reception of its echo is Δt, then the distance of the object is $d = \frac{1}{2} \Delta t \cdot c$. In the practical application of radar, we never take into consideration whether the device is fixed on the ground or installed on a high-speed warship: in either case we use the same light velocity c in our computation, which fact implies the principle of invariance of light speed.

New Law of Velocity Composition

To sum up, we have obtained two principles concerning velocity:
1. In classical physics the following law of velocity composition is used

$$v = v' + u \tag{1}$$

2. As for the velocity of light, we have
   c = an invariable quantity.

These two principles are "contradictory" to each other, yet both are correct. Obviously a more consistent theory is required to unify these two principles, and this necessitates the development of the classical law of velocity composition into a new law of velocity composition that may contain in itself the constancy in the velocity of light. It is the formula of velocity composition of the special theory of relativity that has satisfied the above requirement. The formula reads

$$v = \frac{v' + u}{1 + \frac{v'u}{c^2}} \tag{2}$$

where all the symbols have the same meaning as in formula (1).

How formula (2) is obtained, we shall discuss by and by. Here let us study its physical import first. Under ordinary conditions, the velocities of all objects are far smaller than that of light, and for which reason we can treat the latter as infinity. Let $c \to \infty$, formula (2) will reduce to formula (1), i.e. (2) embodies the truth of formula (1). On the other hand, if the object under investigation is light, then the velocity of light relative to K' is c, i.e. $v'=c$. Substitute this in formula (2), we shall have $v=c$. That is to say, no matter how great the relative velocity u between K and K' is, the velocity of light measured by either of them is invariably c. Therefore formula (2) also contains the truth of the invariance of the light speed.

The Velocity of Light is a Limit

Now let us make a further comparison between the classical formula (1) and the relativistic formula (2).

In the previous chapter we have discussed the throwing action of the javelin athlete: his run-up is meant to increase the velocity of the javelin relative to the ground. Now suppose that his running velocity approaches the speed of light: can that make the javelin speed exceed the latter? According to formula (1) this is possible. Now let us say that the velocity of the athlete is $u=0.9c$ and the velocity of the javelin relative to the athlete is also $v'=0.9$ c, (both are smaller than c), then the javelin velocity relative to the ground is $v = v' + u = 1.8$ c (which exceeds the velocity of light). In truth, in classical mechanics, the velocity composition law meets no upper limit. Employing (1) repeatedly, we can obtain velocity of any magnitude by adding up a number of smaller velocities.

With the transition to the relativistic physics, the conclusion is altered accordingly. According to formula (2), the velocity of the javelin relative to the ground in the above example should be

$$v = \frac{0.9 + 0.9}{1 + 0.9 \times 0.9} c = 0.995 \text{ c}$$

which does not exceed the velocity of light. That is to say, no matter which frame of reference is adopted, the velocity of the javelin can never exceed light. To speak in a general sense, we conclude that the ultimate result of the composition of a number of velocities smaller than the velocity of light can never exceed the velocity of light.

Thus the velocity of light turns out to be the limiting velocity of all moving objects—another aspect of the absoluteness of the velocity of light.

On Superluminal Speed

A few annotations should be given to the statement that the speed of light is a limit.

An incorrect idea is that the speed of light is the limit of all possible speeds.

No. The speed of light is the limiting speed of an object in motion, or the limiting speed of energy transference. If we discuss speed in a general sense without being restricted by this particular condition, it is not difficult to find superluminal speed phenomena in physics.

Here let us cite a very common example. On a festival night when the beam of the search-light reaches the high-altitude cloud, an illuminated spot will then be seen thereon as the result of reflection. As the search-light device on the ground is

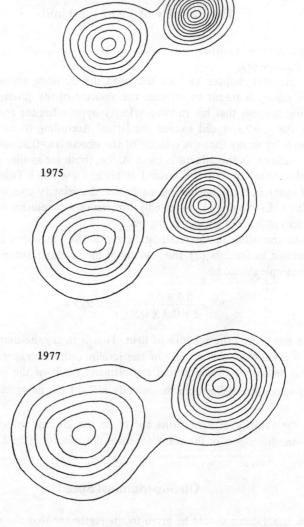

Fig. 3-4 Quasar 3C345 has two radiosources receding from each other since the 70's at superlight speed

slowly rotating, the bright spot meanwhile will be moving at a very high speed. If the cloud is sufficiently high, the speed of the bright spot can exceed that of light. Since along the track of the bright moving spot no energy is being transferred, its speed should not be limited by that of light.

The search-light example not only serves to explain a principle, but its application may bring forth some real value. Since the seventies, the resolving power of radio telescopes has greatly improved. Now that the very long baseline interference device has been in use, its resolving power can be raised to that extent as can make the operator at Lasa see clearly a stamp at Harbin.[1] The application of the new technique enables us to find out that many quasars contain two symmetrical dauble radiosources (see Fig. 3—4). The most interesting thing is that in some of the quasistellar objects these radiosources are separating from each other. From the rate of increase of the distance between them their separating velocity can be calculated. As regards such quasars as 3C345, 3C273, 3C279, their separating velocities all surpass that of light, in one case even amounting to ten times as great as the latter.

The mechanism of one of the models employed to explain this superluminal speed is as follows: when the two streams of particles (corresponding to the searchlight beam) ejected in two opposite directions by the mother body at the centre of the quasar reach the interstellar media (corresponding to the high-altitude cloud), the latter, being stimulated, become radioactive (corresponding to the illuminated spot). Therefore, when the mother body at the centre vacillates just a little, the radioactive regions stimulated by the particle streams will move at a tremendous speed. This speed is not limited by that of light, i.e. superluminal speed in this case is permissible.

Of course, the "searchlight" model is only one of the possible explanations given to the motion at superluminal speed. Many other models are also plausible in explaining this phenomenon. Up to now none of them has been generally accepted as the most reasonable explanation. It needs further observation to decide which one is the most justifiable mechanism.

The Measurement of c

Since the speed of light has so many important properties, it has been treated as a basic physical constant.

The first man that tried to measure the speed of light was Galileo. He and his assistant stood several kilometres away from each other at night, each holding in his hand a lamp with a shutter. (People in his time had no knowledge of electricity, not to say electric light). At the instant Galileo opened his lamp, a beam of light went in the direction of his assistant who opened his lamp as soon as he saw it. Galileo meant to measure the interval between the instant he sent out the signal and the instant he saw the return signal his assistant, from which interval he could calculate

---

(1) From Lasa to Harbin the distance is about 3,400 km as the crow flies.

the velocity of light. However, this experiment turned out to be a failure, since the velocity of light is so great, as we know now, that a time difference of 10^{-5} second must be measured if we hope to obtain the velocity of light through this method. At that time to achieve the required precision was entirely out of the question.

The first comparatively correct value of light velocity was obtained through the observation of celestial objects. In 1675 the Danish astronomer O. Roemer noticed that the interval between two successive disappearances of a satellite of Jupiter in the latter's shadow increased or decreased according to whether the earth in its orbital motion was leaving or approaching the planet when these two eclipses of the satellite took place. He believed that this was due to the different distances over which light travelled in the two cases. Based on this judgement Roemer calculeted that $c = 2 \times 10^8$ m/s.

In 1849 light-velocity measurement was carried out for the first time in the laboratory. In that year the French physicist H. Fizeau employed highspeed gears in his measuring work. In 1862 J. Foucault successfully developed another method to measure the velocity of light: he used a highspeed rotating mirror to measure exceed-

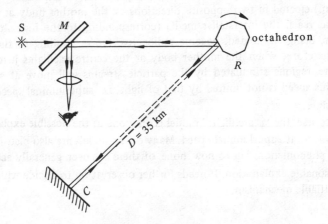

Fig. 3−5 Scheme of the measurement of the velocity of light using a highspeed rotating octahedron

ingly small interval of time. The following description is an improved version of his is experimental scheme. The rotating mirror is a steel octahedron with polished surfaces. When light emitted by S reaches R, the reflected beam travels 35 km, reaches mirror C and is reflected back to the rotating mirror. The whole journey takes $t = \frac{2D}{C}$, during which time the mirror turns a certain angle. Now let us increase the rotating speed in our experiment. As the speed reaches 528 rps, during t the mirror rotates exactly 1/8 cycle when the returned light falls exactly on the next octahedral surface. Through the semi-transparent mirror M, the image of the returned light can be seen through a telescope. By this method the obtained light speed is $c = 299\ 796 \pm 4$ km/s.

The modern method of measurement is first to obtain both the frequency ν and the wavelength λ of a beam of light, and then to substitute them into the formula $c = \nu\lambda$. Since 1973, the following value has been adopted as the speed of light[1]:

$$c = 299\ 792\ 458 \pm 1.2 \text{ m/s}$$

Incidentally, despite the different methods by which the speed is measured, the results are the same—another proof of the constancy of the speed of light.

---

(1) Recently by international agreement the value for the speed of light has become a defined quantity. Correspondingly, the metre is defined as the distance travelled by light in a vacuum in $1/299\ 792\ 458$ s.

CHAPTER IV FROM GALILEO'S PRINCIPLE OF RELATIVITY TO THE SPECIAL THEORY OF RELATIVITY

Salviati's Ship

In the first chapter we have made mention of the fact that classical physics begins with rejecting the Aristotelean conception of space and time.

For a time a heated debate was going on. Those who advocated the Copernican theory maintained that the earth was in motion while others who supported the Aristotel-Ptolemaic system insisted that the earth was at rest. The earth-is-at-rest school opposed the earth-is-in-motion school with this strong argument: if earth is in highspeed motion, why cannot people standing on it feel the motion? This seems to be a real problem.

In 1632 Galileo published his famous *Dialogue Concerning the Two Chief World Systems-Ptolemaic and Copernican,* in which Salviati, a member of the earth-is-in-motion school (Fig. 4–1), gave a thoroughly convincing answer. "Now let us shut you and your friends up in the chief under-deck cabin of a great ship," says Salviati, "wherein a few flies, butterflies and other winged insects are allowed to keep you company. A big bowl filled with water wherein a few fish leisurely swim is also placed in the cabin. From the ceiling hangs a big bottle filled with water which drips,

Fig. 4–1 Salviati's ship

drop by drop, into a big-mouthed urn. When the ship is at anchorage, by careful observation you may notice that these insects are flying at equal speed in different directions in the cabin, the fish are swimming freely in this as in that direction, the water-drops fall into the urn directly below; when you throw anything to your friend, you use the same strength in this direction as in that, provided the distance is equal. When you spring on both feet, you'll make the same distance no matter in which direction you leap. After having carefully studied these things, let the ship sail at any speed, you'll notice, as long as the motion of the ship is uniform and the ship does not roll and pitch, all the above-mentioned phenomena will remain completely unchanged. You can never decide by any of these phenomena whether the ship is in motion or at rest. Even when the ship moves at a comparatively great speed, just as before in leaping you'll cover the same length on the floor: the distance you make in jumping astern will not be any greater than that if you jump ahead. During your jumping flight, the floor of the ship moves in the direction opposite to that in which you jump. No matter what object you throw to your friend, no matter whether he stands at the bow or at the stern, as long as you stand facing him, you'll use the same strength. The waterdrops will fall into the urn below, not a single drop will fly to the stern, though during its falling the ship has already covered quite a few spans[1]. The fish swim towards the front part of the bowl with no more effort than towards its back part: they leisurely swim to any part of the rim seeking food. Last of all, the butterflies and the flies fly freely in the air, showing neither the tendency of gathering at the stern, nor any sign of fatigue caused by their prolonged stay in the air and their effort to catch up with the motion of the ship from which they now hold aloof."

Salviati's ship illustrates an important truth: from each and all phenomena that can be witnessed in the ship, one can never ascertain whether the ship is in motion or at rest. This assertion is now called the Galilean principle of relativity.

In modern language, Salviati's ship is nothing other than a so-called inertial frame of reference. Any ship in uniform motion without pitch and roll may serve as an inertial frame. All the phenomena that can be seen in one inertial system can also be seen in another without the slightest difference. In other words, all the inertial frames of reference claim equal right: they are all equivalent. We can never decide which inertial frame of reference is at absolute rest and which is in absolute motion.

Galileo's principle of relativity not only knocks the bottom out of all the arguments the earth-is-at-rest school holds against the earth-is-in-motion school, but also reveals the fallacy of the concept of absolute space (at least in the field of inertial motion). Therefore, in the transition from classical mechanics to the theory of relativity, many of the classical concepts should be altered with the only exception of Galileo's principle of relativity which needs not to be revised in the least. Moreover, this principle of Galileo's has become one of the two basic principles of the theory

---

(1) An ancient unit of length indicating the distance between the thumb and the little finger of a spread hand, usually taken as 9 inches.

the special relativity[1].

Two Principles of Special Theory of Relativity

In 1905, Einstein published his paper *On the Electrodynamics of Moving Objects* in which he laid the foundations of his special theory of relativity. Concerning the basic principles of the special theory of relativity, he wrote:

"The following considerations are based on the principle of relativity and the principle of the constancy of the velocity of light, which two principles are defined as follows:

"1. All the laws that govern the changes of the states of the physical systems have nothing to do with which of the systems of coordinates that are in uniform motion relative to each other is employed in describing the changes of the states.

"2. Any light beam travels with a constant velocity c in the 'still' coordinate system, whether the object that emits the light is at rest or in motion."

The first one is the principle of relativity, whereas the second one is the constancy of the velocity of light. The complete special theory of relativity is based on these two basic principles.

Einstein's philosophy is that nature is simple and harmonious. His theory always has this appealing feature: it looks simple but it is deep. The special theory of relativity is a system marked with this feature. These two basic principles seem to be such "simple facts" as can be accepted without difficulty, yet their deductions have radically changed the foundations of the Newtonian physics.

Now let us begin the deduction as follows.

Simultaneity Is Relative

In the first chapter we have already mentioned the relativity and the absoluteness of the concept of simultaneity. Here let us discuss them more carefully.

By the simultaneity of two events we mean that the positions of these two events may be different, yet they happen at the same time. For example, anytime the radio station broadcasts the time signal, many people at different places will set their watch or clock by it. Hence we say that the watch setting by people at different places happen simultaneously. Yet a more careful analysis shows that this assertion is

---

(1) In many textbooks Galileo's principle of relativity is called mechanical theory of relativity, to be distinguished from what is called the principle of relativity in the special theory of relativity. The difference is, the former insists the equivalence of all the inertial frames of reference in surveying any mechanical phenomenon whereas the latter furthers the generalization by stating that the surveying of any physical phenomenon is equivalent in all the inertial frames of reference. In fact, this distinction does not fall in completely with the historical facts, since "Salviati" unequivocally refers to "any phenomenon", not only the mechanical phenomenon.

not completely correct. Since it takes the signal a certain length of time to reach the radio-receiver, the time different radio-receivers at different places receive it will not be exactly the same: the farther the receiver is away from the station, the longer time it takes the signal to reach the receiver. Since the speed of the radio wave is so great, such small disparity ensuing from setting watch in this manner is negligible. In our daily life imprecision as such will cause us no trouble at all.

Yet when we discuss the principle of a problem, even the slightest imprecision cannot be tolerated. To be exact, only when two clocks are placed at equal distance from the radio station can they receive the signal at the same time.

Now let two clocks be placed at A and B separately, as shown in Fig. 4–2.

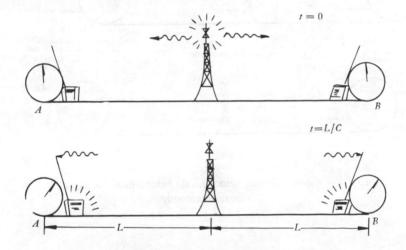

Fig. 4–2 In the same inertial system people can set clocks at A and B by the radio time signal.

Their distances to the radio station are the same, i.e. L. Now let us say at $t=0$ the radio station emits the signal, then at $t=L/c$ the signal reaches A and B simultaneously. Through this process, the radio station can help people to set all the clocks placed at different points of the same inertial system.

Now if some one stands on another inertial system K' which moves to the left with a velocity v relative to K (Fig. 4–3), in his eyes the radio station, A and B all move with velocity v to the right, while the distances of A and B to the station remain equal, let us say L'. Due to the constancy of the velocity of light, relative to K', the velocity of the signal remains c. Since A has a velocity v to the right, in K's eye, the relative velocity between the A-ward moving signal and A is $c+v$[1]. We can

---

(1) Here we come across the limiting velocity of light again. c+v exceeds the light velocity, yet this does not contradict the limiting velocity of light. By the constancy of the velocity of light we mean the velocity remains unchanged with respect to the observer. Here c+v is the relative velocity between a beam of light and another object, as viewed by the observer. This can exceed the velocity of light.

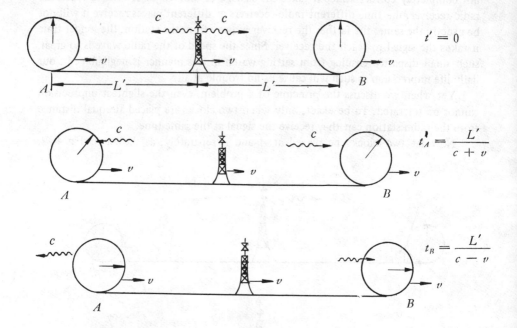

Fig.4–3 With respect to inertial system K', the signal arrives at A and B unsimultaneously

likewise show that the relative velocity between the B-ward moving signal and B is $c - v$. Therefore if the radio-station gives the signal at $t'=0$, then the time A and B receive the signal should respectively be

$$t'_A = \frac{L'}{c+v}, \quad t'_B = \frac{L'}{c-v}$$

Obviously $t'_A \neq t'_B$. That is to say, in the eye of K', the signal arrives at A and B at different time. This proves that "simultaneity" is relative. It depends on which frame of reference is employed. When a different frame of reference is chosen, non-unsimultaneous events may become simultaneous, and vice-versa.

Who Shot First?

According to the special theory of relativity, not only "simultaneity" is relative, sometimes even the sequence of events is also relative. For example, in a

10-metre-long carriage, B stands in the front part of it, A stands in its rear part. When the carriage passes the station at a high speed of 0.6 c, suddenly a man standing on the station sees A start shooting at B, and after 12.5 nano-seconds he sees B shoot back at A. Hence the evidence he gives to the men on the station is: A started the gunfight. But the passengers in the carriage bear contrary witness: they all say B shot first, and after 10 nano-seconds A returned the challenge.

Yet who actually shot first? No absolute answer can be given. In this particular case, even the sequence of events is relative. If the carriage is used as the reference frame, B shot first, A shot next. If the station is used as the reference frame, A shot first, B shot next.

Fig. 4–4 Who shot first?

Cause and Effect

The previous example will certainly bring doubt to some readers' mind. If the sequence of events is relative, is it possible that in a certain reference frame people can witness the death of a man previous to his birth, or the arrival of a train before its start? Speaking more generally, since cause always precedes effect, will the alteration of the sequence of events bring about such confusion as effect preceding cause?

To clarify this problem, let us see the following diagram (Fig.4–5). The horizontal axis represents the space coordinate x, the perpendicular axis represents the time coordinate t. If the event is located at the origin (i.e. $x=t=0$ event), then the worldlines of the light emitted by or returned to it are two lines inclined at 45° (if we let the velocity of light $c=1$). These lines cut the whole plane into 4 cone-shaped areas. Now the origin event 0 can be connected with any event in the areas I and II, but cannot possibly be connected with any event in the areas III and IV by means of a signal whose speed is equal or less than that of light.

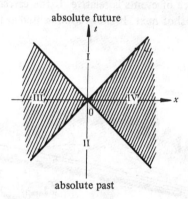

Fig. 4—5 The light-cone diagram. Event 0 cannot form cause-effect relations with any event in the areas III and IV, but can form cause-effect relations with events in I and II

Since the speed of light is the limiting speed, event 0 cannot be related with any event in the areas III and IV through any possible signal. Any two events that cannot be related through any signal cannot be cause and effect to each other. Therefore, so far as these events are concerned, the relativity of before-and-after has nothing to do with the relation of cause and effect. On the contrary, 0 can be related with any event in the areas I and II by means of signal, therefore the sequence of before-and-after of these events claims no relativity, thus in no case contradicting the relation of cause and effect.

Relative to event 0, area II is the absolute past whereas area I is the absolute future. This before-and-after sequence cannot be altered by the selection of reference frame: it is absolute. Therefore the special theory of relativity satisfies the requirement of cause-and-effect relation.

In Newtonian physics, we are not certain what is the prerequisite of the cause-and-effect relation between two events. Einstein's physics demonstrates this: the prerequisite of the cause-and-effect relation of two events is that they can be related through signals whose speed is equal to or less than light. Now Let us return to the previous discussion of the gunfight between A and B. Since A and B do not satisfy this necessary condition (in the duration of a little more than 10 nano-seconds light cannot travel as far as 10 metres), the before-and-after sequence of the shooting action of A and B is relative.

Here we see again the importance of the velocity c. It is exactly the constancy of the light velocity that at once makes possible the establishment of the cause-and-effect relation and prevents us from seeing any inversed order of cause and effect.

Now let us sum up briefly the various differences ensuing from the transition from classical mechanics to the theory of relativity. In the following table "absolute" means no change occurs with the choice of reference frame, "relative" means change ensues with the choice of reference frame.

	classical mechanics	special theory of relativity
velocity of light	relative	absolute
simultaneity	absolute	relative
order of two events which cannot be physically related	absolute	relative
order of two events which can be physically related	absolute	absolute

CHAPTER V RELATIVITY AND ABSOLUTENESS OF ROD AND CLOCK

Space and Time in Newtonian Space-time Concept

To the table of classification of relativity and absoluteness given at the end of the previous chapter, we may add new contents step by step. Newton's space-time also contains another two absolute concepts, i.e, the length of rod and the duration of time.

When a man sees from his own watch the passage of one minute, he will naturally think that all the watches and clocks in the world have ticked away the same duration of time, disregarding the different moving conditions in which they are placed. That is the absoluteness of the time duration.

Similarly, the length of a straight rod, if it measures one foot in a certain frame of reference, will measure one foot in any other frame of reference. That is the absoluteness of the length of rod.

The absoluteness of time-duration and the length of rod plays an important role in Newton's space-time concept, but in the theory of relativity it has turned relative.

The Moving Clock's Time Slows Down

In Chapter II we have already said that any device by which time can be measured is a kind of a clock. Making use of the constancy of light velocity, we can also build a kind of radar-clock, the construction of which is sketched in Fig. 5–1.

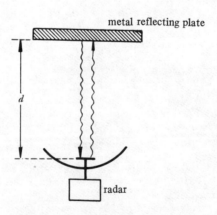

Fig. 5–1 The structure of radar-clock

The device consists of an emitter and a reflecting metal plate, the distance between them being d. After being reflected by the plate, the emitted signal is received by the emitter antenna again. Since the round trip of the signal is $2d$ in distance, and the velocity of the signal is c, time taken by the round trip is $T=2d/c$.

How to use the radar-clock to measure time? If during a complete process, the radar signal has made 5 round trips, then we say that the process takes $5T$. If the signal has completed 3 round trips, then the process takes $3T$. That is to say that we take the duration of one round trip of the signal as the unit whereby to measure the duration of time.

Now let us say that A and B each have a radar-clock. When A and B are at rest relative to each other, let them synchronize their clocks. After that let them be in relative motion. A and his clock move leftward, while B and his clock move rightward. Then what will they see separately?

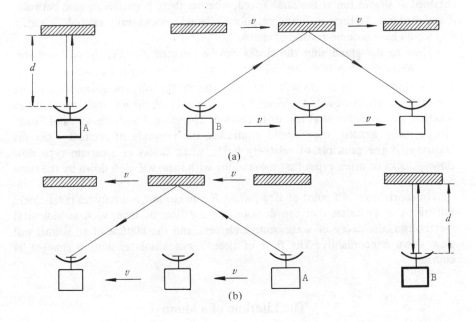

Fig. 5-2 Moving clock slows down

Let us discuss A first. With respect to A himself, A is at rest, his clock is also at rest. In his eyes, his clock remains as before: it has not changed in the least (just like things in Salviati's ship). In the meantime A sees B and his clock move rightward. In the emission-reflection-reception process of B's clock, the antenna and the reflecting plate are moving continuously, therefore the signal describes oblique lines (Fig. 5-2 (a)). Therefore in A's eye, the round trip of B's clock measures a length greater than $2d$. Yet due to the constancy of the velocity of light, both the signal of A's clock

37

and the signal of B's clock have the same speed c. Therefore what A sees is, when A's clock indicates the passage of a unit time, the signal of B's clock has not yet completed a round trip. A concludes: "B's clock is slower than mine."

Now if we let B view the entire drama, then everything is reversed. B will consider himself at rest, while A's clock moving leftward (Fig.5–2 (b)). The distance the signal of B's clock travels is $2d$, while the round trip of the signal of A' clock, its route being oblique lines, is greater than $2d$. Therefore B concludes: "A's clock is slower than mine."

Yet which of them is right? Both are right. Their conclusions, though contradictory in appearance, are actually in agreement. Their common conclusion is: A moving clock slows. From A's point of view, B is in motion. From B's point of view, A is in motion. Therefore each of them witnessed the slowing down of the other's clock.

Some may doubt the universality of the conclusion. They think the use of radar-clock is to blame. In their opinion such "good" clocks can sooner or later be obtained as should run at the same speed, whether there is relative motion between A and B or not. To their astonishment, if such "good" clocks really existed, Salviati's ship would have become a world of chaos.

Then in the grand ship the clocks can be assorted into the "good" and the "bad". When the ship is at rest, all the clocks agree in speed. But when the ship begins to move, some clocks will speed up, other clocks will slow down. If that were the case, we would be able to decide by the disparity of these two classes of clocks whether Salviati's ship is at rest or in motion. Therefore if such "good" and "bad" clocks really existed, they would contradict the principle of relativity. On the contrary, if the principle of relativity holds, when clocks of a certain type slow down, clocks of other types that move along with them will slow down by the same amount.

In short, from A's point of view, when B is in motion, not only his radar-clock, but all other processes that can demonstrate the flow of time, such as biological metabolism, the decay of a radioactive element and the lifetime of an animal will slow down concordantly. The flow of time is not absolute: motion changes its course.

The Lifetime of a Muon

Lifetime can also serve as a kind of "clock". The span of a generation as often heard in our daily speech is the duration of time measured by life as a clock. Therefore life is not absolute either. The lifetime of a thing, viewed from different frames of reference, should be different accordingly. In fact, things are really so.

There is a certain particle called muon (μ-meson). It is unstable, and its lifetime is very short: from birth to decay it lives only 2 microseconds (2×10^{-6} s). Therefore even if a muon travels at the speed of light, it can cover a distance of $2 \times 10^{-6} \times c \approx 600$ metres only. Yet the study of cosmic rays shows that the muon born in the high-altitude firmament can even reach the earth: the distance it has covered by far exceeds 600 metres. How come everything like that? The phenomenon of the slowing-down of a moving clock may help us to solve the riddle.

In high-speed motion, a life "clock" slows down in the same way as all other clocks do. Therefore, the lifetime of a muon in high-speed motion by far exceeds 2×10^{-6} s, and its distance of flight by far exceeds 600 metres.

Fig. 5–3 demonstrates the relation between the velocity of the moving object and the slowing-down of time. The horizontal axis indicates the velocity of the

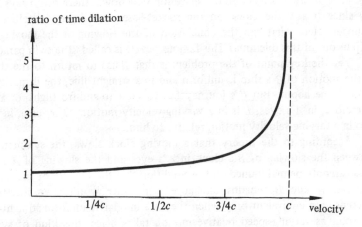

Fig. 5–3 The relation between the velocity of the moving object and time dilation

moving object, the perpendicular axis indicates when the moving clock shows the lapse of 1 second, how long time has passed according to the clock at rest. For example, when 1 second has passed according to the clock moving with the velocity of 0.6 c, 1.25 seconds have passed according to the clock at rest. From the diagram, the reader can see clearly that, only when the velocity is very close to that of light, in the eyes of the motionless observer the effect of the lengthening of life becomes conspicuous. Only when the velocity literally equals that of light, the motionless observer will witness the infinite lengthening of the life span of the one in motion. Here again we see the limiting velocity of light.

The Twin Paradox

A man, just like a muon, has a definite lifetime, let us say 100 years at most. If we do not take into consideration the slowing of time of a moving clock, even if a man sails in a light-speed rocket, his journey will not exceed the bounds of 100 light years: he can never reach the remote stars or galaxies. Yet in fact the lifetime of the passenger in the light-speed rocket will be greatly lengthened in the eyes of the people on earth, therefore his journey can by far exceed 100 light-years. On the other hand, in the eyes of the passenger in the rocket, the earth is also flying away from him. Therefore, in his eyes the life of the people on earth is also lengthened. When the distance between the earth and the rocket exceeds 100 light years his brothers on the earth are still living.

Here we meet another difficulty.

Let us suppose that A and B are twin brothers. They have planned a highspeed spaceship journey to test the validity of the special theory of relativity. A will stay at the rocket-base, while B will make a round trip in the outer space. When the spaceship returns, is A younger than B, or B younger than A? Here are two answers: (1) Since A saw B's clock in the spaceship slow down, therefore A says B is younger; (2) since B saw the clock on the rocket-base slow down, therefore B says A is younger. Here what can the conclusion of the slowing of the moving clock do to help us out of this dilemma? This famous puzzle is called "the twin paradox".

Yet the key point of the problem is that B has to return to the starting point. If the motion of B's ship is uniform and in a straight line, the disparity of age will never come about. But B's journey has to be a to-and-fro flight, or a great circle. Therefore, in A's eyes, B is in a varying-velocity motion. Of course, in B's eyes A is also in a varying-velocity motion relative to him.

According to the theory that a moving clock slows, the symmetrical equality between the slowing of B's clock in A's eyes and the slowing of A's clock in B's eyes can only be maintained on the condition that the velocity of the relative motion between A and B remains unchanged. In other words, two inertial frames of reference are equivalent only when they are in relative uniform straight-line motion. As soon as varying-speed relative motion takes place, this kind of symmetry will be spoilt.

But don't forget that both A and B are living in the universe. Around them there are numerous celestial objects. Therefore the twin paradox consists of three factors: A, B and the surrounding world. As long as A stays on the base, the innumerable celestial objects are not in varying speed motion. In his eyes only B is moving with varying velocity. Yet in B's eyes things are different. With respect to him not only A is moving with varying velocity but the whole universe is moving likewise. On one side is the whole universe, on the other side only the spaceship—that is an obvious asymmetry. Here the dilemma ensuing from symmetry does not really exist. Then which of them is younger?

In the year 1966, a twin-paradox experiment was actually made to decide which of them would enjoy a lengthened longevity. The experiment brought the pure theoretical controversy to an end, once for all. But this time the traveller is not a man, but a muon. The route of the journey does not lie in the outer space but along a circle 14 metres in diameter. The muon starts at a certain point, and journeying along the circular track, returns to the starting point, in the same manner that B completed his space journey. The result of the experiment is: the muon after the journey is really younger than his brothers at home. It seems that we can draw such a conclusion: the farther one moves with varying velocity relative to the whole universe, the longer he lives.

Fig. 5-4 The twin paradox

The Moving Rod Contracts

Now let us turn to the relativity of the length of rod.

In 1893, to explain the Michelson-Morley experiment, G. Fitzgerald first and H. Lorents afterwards put forward the hypothesis that the length of any object contracts in the direction of its motion. The phenomenon was later called the Lorentz-Fitzgerald contraction. According to Fitzgerald's quantitative description of the contraction, a rocket flying at the speed of 11 metres per second only contracts two billionth parts of its length. But in high-speed motion the contraction of a rod is remarkable. Fig. 5—5 shows the diminishing of the length of a one-metre rod in its motion. When the velocity reaches one half of the light velocity, the contraction amounts to 15 per cent. When the velocity reaches 260 000 km/s, the contraction is 50 per cent, i.e. a one-metre rod now becomes 50 centimetres.

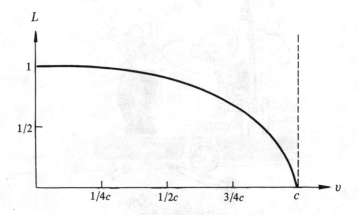

Fig. 5—5 Lorentz-Fitzgerald contraction

In the special theory of relativity, the length of a rod is also relative (determined by the choice of the frame of reference). The length of a rod changes in the same manner as Lorents-Fitzgerald previously stated in their supposition. Here an additional explanation will be made about how the length is measured. If a rod is motionless relative to a certain frame of reference, then from the difference of the space coordinates of the two ends of the rod can the length of the rod be obtained.

When the rod is in motion relative to a frame of reference, we can measure the length of the rod in the following way. At a given time let two men take photos of the rod, one photographing its front end, the other photographing its rear end. Since the two photos are taken simultaneously, by ascertaining the difference of the space coordinates of the two pictures through comparing them, we can obtain the length of the moving rod. Here the key words are "two photos taken simultaneously." As we know in the relativistic concept of space and time, "simultaneity" is relative: it depends on the choice of the reference frame. Therefore, with respect to different frames of reference, photos are taken in accordance with different "simultaneities". Hence it is not difficult to predict that different results will ensue from the measurement.

The length contraction just like the time dilation is symmetric, i.e. if between A and B there is relative motion, then, just as A witnesses the contraction of B's rod, so will B witness the similar contraction of A's rod. This conclusion shows that spatial scales are not absolute, but relative.

Mr. Tompkins' Mistake

Mr. Tompkins is the hero in *Mr. Tompkins in Paperback*[1] by G. Gamow. The author of the book says that, when Mr. Tompkins came to this wonder city, very easily he saw, due to the unusually low limiting velocity (corresponding to the light velocity in the real world), all kinds of relativistic effects. Mr. Tompkins proclaimed, when he rode his bicycle at a high speed, that he discovered the city distorted into what is shown in Fig. 5–6.

Fig. 5–6 Tompkins' experience

Tompkins' experience has been acknowledged as correct by the physicists in the recent decades. All believe, if we can move at a speed near that of light, we shall, like Tompkins, see a flattened world. From the relativistic effect of contraction of a moving rod, we seem to have come to this conclusion very naturally.

But the conclusion is wrong. The contraction of the moving rod cannot prove that the world as Tompkins saw it was a flattened one. The key point is that, the

---

(1) Cambridge University Press, 1965.

contraction of a rod is obtained in accordance with "photos taken simultaneously." Yet the very "witness" of Tompkins does not satisfy this requirement. When an eye "sees" an object, it means the photons emitted by all parts of the object enter the eye simultaneously whereby an image is formed. Hence it follows that these photons cannot have been emitted at the same time, since different parts of the object lie at different distances from the eye. The point farther away from the observer has to emit its photons earlier; the nearer point, on the other hand, later. This requirement runs directly against that of "simultaneity" in the measurement of the rod length.

Therefore we can never see the scene as Mr, Tompkins described. Yet what scene is to be seen?

Now let us consider a cube of 1 metre each side. When this cube is at rest, an observer at a comparatively great distance to the cube, perpendicular to bc, can only

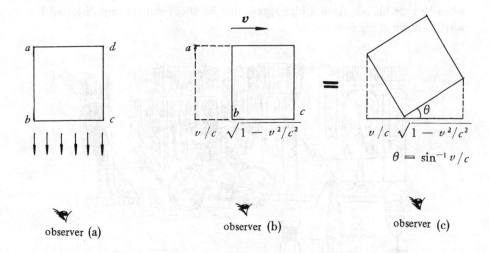

Fig. 5−7 When the cube moves with velocity v, the observer will see a rotated cube, the angle of rotation being $\theta = \sin^{-1} v/c$

see the bc face of the cube. The light emitted by point a cannot be seen by him (Fig. 5−7 (a)). When the cube moves in bc direction with a high velocity v, it contracts in the bc direction, its length becoming $\sqrt{1-v^2/c^2}$ (Fig.5−7 (b)). But now the observer begins to receive the light emitted from point a. Since the light emitted by a and the light emitted by b and c reach the observer's eye simultaneously, point a's emission of light seen by the observer is $1/c$ sec earlier than the emission of light of the bc side. But in the duration of $1/c$ sec the cube has moved a distance of v/c. Therefore the observer now can see the ab side. In a word, what object the observer actually sees is none other than a rotated cube, the angle of rotation being $\theta = \sin^{-1} v/c$ (Fig. 5−7 (c)).

From this example we can see that the effect of the rod contraction affords to us not a flattened, but a rotated, view of the object. We can generally prove, when an object of any shape moves with velocity v, the view of the object, in the "eyes" of the observer, has only obtained a slight rotation relative to its form at rest: it never becomes flattened.

Lorentz Transformation

The previous discussions have touched on many aspects of the theory of relativity. In them we find one thing in common: we have to consider the position and time of the same event from two different frames of reference. Whether with respect to the problem of simultaneity, or to the problem of rod contraction and time dilation, we have to ascertain the position and time of the event relative to the frame of reference K on the one hand, and to ascertain the position and time of the event relative to the frame of reference K' on the other, while between K and K' there is relative uniform motion in the meantime. Therefore the essence of these problems is that we need to find out the relations between the space and time coordinates of each event relative to frame K and its space and time coordinates relative to frame K'.

Now let us say that the space coordinates of an event are $x, y, z,$ and the time coordinate of it is t with respect to frame K. Then what are the space coordinates x', y', z' and the time coordinates t' with respect to frame K'? To simplify the problem, let us suppose that between K' and K there is only relative motion in the direction of the x-axis, the velocity of the motion being v (Fig. 5-8). According to

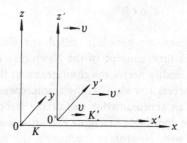

Fig. 5-8 Two inertial frames K and K' are in relative uniform motion

the principle of the constancy of light velocity and the principle of relativity, we can obtain the following transformation relations between coordinates x, y, z, t and coordinates $x', y', z' t'$:

$$x' = \frac{x - vt}{\sqrt{1 - v^2/c^2}}$$

$$y' = y$$
$$z' = z$$
$$t' = \frac{t - \frac{v}{c^2}x}{\sqrt{1-v^2/c^2}}$$

which equations are the well-known Lorentz transformation.

Lorentz transformation equations are the core of kinematics of the special theory of relativity. From these equations we can naturally derive all the quantitative relations of the relativistic effects we have previously discussed. For example, if the length of a rod at rest is L_0, then when it moves with a velocity v relative to the observer, its length becomes $L = L_0 \sqrt{1-v^2/c^2}$. Similarly, when a clock moving with velocity v relative to the observer indicates the passage of $\Delta t'$, the other clock at rest will indicate a lapse of $\Delta t = \frac{\Delta t'}{\sqrt{1-v^2/c^2}}$ Fig.5–3 and Fig.5–4 are designed in accordance with these equations. As for the Lorentz transformation, let us say something more. Under ordinary conditions, the velocity of a moving object is always by far less than the velocity of light. Therefore, if we regard the light velocity c as infinite, then the above equations will reduce to

$$x' = x - vt$$
$$y' = y$$
$$z' = z$$
$$t' = t$$

This set of equations are generally called the Galilean transformation—the foundation of the space-time concept of the Newtonian mechanics. From Galilean transformation we can readily derive the absoluteness of the interval of time and the length of an object, whereas $t'=t$ signifies the absoluteness of simultaneity. Galilean transformation is only an approximation to Lorentz transformation. Lorentz transformation equations are applicable to a more extensive scope of phenomena. That is to say, in comparison with Newtonian mechanics, the special theory of relativity is a more correct description of nature.

CHAPTER VI   DYNAMIC PROBLEMS

Aristotelean Dynamics

Dynamics studies the cause of motion of an object. Stated simply, it answers why an object moves, why it moves in this, and not in that manner, etc.

To answer these questions from everyday experience seems not very difficult. When we walk, we have to exert ourselves. The motion of a waggon is caused by the pulling of a horse. An airplane flies because of the driving of its engine. These observations create in our mind the conception that the cause of motion is force, that motion cannot take place without force, and that force is the decisive factor of motion. Basically the conception is correct, but the next question is: In what manner does force determine the character of motion of an object?

Aristotle gave this answer to the question: force determines the velocity of a moving object. Indeed, if we want the waggon to go faster, we need more horses to pull it, or replace the horses by stronger ones. Therefore the greater the force, the higher the velocity; the less the force, the lower the velocity. When no force is applied, the velocity comes to zero (i.e. the object becomes motionless). This is Aristotle's dynamical law.

The Moving Object Moves Forever

Since Aristotle's dynamical law can explain many everyday phenomena, albeit superficially, in medieval Europe both the Church and the secular world regarded it as a credo.

The first man that set to refute the law was again Galileo. He first noticed that each object has a definite inertia. For example, after the horses cease pulling a fast moving waggon, it takes the latter some time to come to a stop.

This phenomenon cannot be explained by Aristotle's mechanics. According to Aristotle's mechanics, as soon as the pulling force ceases to act on the waggon (i.e. the horses cease to exert themselves), the velocity of the moving object would immediately come to zero (i.e. the waggon should stop all of a sudden). Therefore from the time the horses stop pulling the waggon to the time the waggon becomes completely motionless, the cause of the motion of the waggon cannot be ascribed to the external pulling force, but to something else. And this something else is none other than the inertia of the object—the innate inclination of an object to keep itself in its original state of motion.

But how long can the motion subsist on inertia? Since the waggon comes to a halt so soon, it seems the motion ensuing from inertia can only go on for a limited period of time. Yet this is only an incomplete modification of Aristotle's mechanics.

Galileo did not stop here. He analyzed an ideal experiment. The experimental device consists of a smooth sloping plane with a little ball rolling downward thereon. The smaller the angle of inclination (or the longer the sloping plane) is, the less gravitational pull is exerted on the ball. When the angle of inclination comes to zero (i.e. the plane lies in a horizontal position, its length becoming infinite), the horizon-

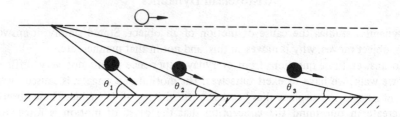

Fig. 6—1 An ideal experiment that can prove what is moving will move on forever

tal pull of gravitation comes to zero. On the sloping plane (the angle of inclination being non-zero), as long as the surface is very smooth, the little ball can always roll down. When the little ball starts rolling on the sloping plane, let the angle of inclination approach zero, then the little ball, though no pulling force is being exerted on it now, will nevertheless travel to an infinite distance. That is to say, the little ball can move everlastingly without any external force acting on it. Here the motion of the little ball only relies on inertia. Therefore inertia can keep an object moving perpetually. The reason why the waggon comes to a halt within a definite period of time is because the ground offers resistance to the waggon. If the ground is as smooth as the sloping surface in the ideal experiment, the waggon will continue its motion everlastingly.

This is the law of inertia in the Galilean mechanics, or the famous theory of "what is moving will move on forever". In more accurate language the theory can be expressed as follows: an object, when no external force is acting on it, will forever maintain its own state of motion: the velocity of its motion can neither increase nor decrease.

This is a complete negation of Aristotle's mechanics that velocity is determined by force. The mechanics based on the inertial law insists that an object, when no external force is acting on it, can have any velocity and maintain its constancy eternally.

In what manner, then, does the force affect the motion of the object? Galileo had not answered this question.

Newton's Mechanical Laws

Yet Newton had answered the above question.

Newton's conception was: the effect of force is not to determine the velocity of a moving object, but to change it. The greater the force, the greater the rate of change; the less the force, the less the rate of change. When no force is applied to the object, the velocity will remain the same. The last point is none other than Galileo's inertial law.

Newton introduced the concept of acceleration for describing the rate of change of velocity. His mechanical law says: the force applied to an object is proportional to the acceleration of the said object, the proportionality factor being termed as the inertial mass of the object. If we use a formula to describe it, it will be

$$ma = f$$

wherein f represents the external force applied to it, a is the acceleration of the body, m is the inertial mass of the object.

Therefore, according to Newton's mechanics, with respect to a definite object (i.e. a definite m), acceleration is directly proportional to the external force; with respect to a definite external force (i.e. a definite f), the greater the inertial mass, the less the acceleration.

Following is a table demonstrating the human knowledge of the kinetic laws from Aristotle to Newton.

	relation between force and motion	equation
Aristotle	force determines velocity	v is the function of f
Galileo	inertia maintains uniform motion	When $f=0$, v is constant
Newton	force determines acceleration	$a=f/m$

Up to the development of the theory of relativity, Newton's mechanics was said to be ever-victorious. Only after the development of the theory of relativity were some new contents added to the above table.

Contradiction between Newtonian Mechanics and the Limiting Velocity of Light

According to Newtonian mechanics, a definite force gives to the object a definite acceleration. That is to say, the object in any unit time will acquire a definite amount of increase (or decrease) of velocity. We may use the following diagram to demonstrate the relation. In the diagram the horizontal axis represents time and the perpendicular axis represents velocity. Under the influence of a constant external force, the velocity will increase linearly. Therefore, provided that the application of the

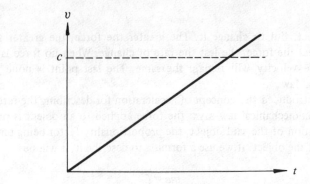

Fig. 6−2 According to the Newtonian mechanics, under the influence of a constant force, the velocity of the object will increase linearly

external force lasts for a sufficiently long time, the velocity of the object will eventually surpass that of light (the broken line). Therefore the Newtonian laws are not compatible with the relativistic concept of space and time. " A definite force determines a definite acceleration" must be incorrect in the theory of relativity.

Inertial Mass Changes With Velocity

Obviously, as required by the limiting velocity of light, the dynamical laws must have the following nature: the nearer the velocity of an object being acted by an external force approaches that of light, the smaller the acceleration produced by the external force will be. When the velocity of the object approaches that of light, the external force applied to it will not produce any acceleration. This will ensure that, no matter how long the external force will be applied to the object, it will not raise the velocity of the object above that of light. If we draw a velocity-time diagram as we did previously, then the velocity of the object being acted on by a constant external force will change in such a manner as shown in Fig. 6−3. In the beginning the acceleration is the same as that obtained from the computation on the basis of the Newtonian mechanics, yet it gradually decreases until finally the velocity steadily approaches c.

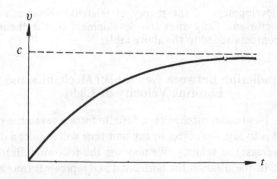

Fig. 6−3 According to the theory of relativity, the velocity of the object being acted on by a constant external force, will change less and less and in the end will steadily approach the light velocity c

If we define the inertial mass as the proportionality constant of the external force divided by acceleration, i.e.

$$m = \frac{f}{a}$$

then, in relativistic mechanics, inertial mass is not a constant but a quantity determined by velocity. The greater the velocity, the greater the inertial mass. When the velocity approaches that of light, the inertial mass approaches infinity. It is only when the velocity is near zero that the inertial mass will be the same as that in Newtonian mechanics. In the special theory of relativity, the quantitative relation is

$$m = \frac{m_0}{\sqrt{1-v^2/c^2}}$$

where v is the velocity of the object and m_0 is the mass when the object is at rest. Fig.6–4 shows the relation between the inertial mass and the velocity. From the illustration it can be seen that, when $v \approx c$, m changes conspicuously with v.

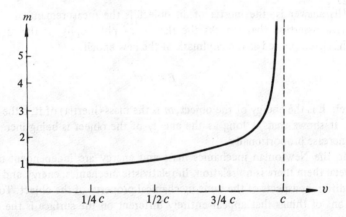

Fig. 6–4 Mass increases with velocity

Inertia = Energia – a Foundation Stone of the New Epoch

Now let us analyze the aforementioned problem from energy consideration.

In Newtonian mechanics, we know, if a force f is applied to an object, then generally speaking, the force will do work on the object, and the work will be transformed into the kinetic energy of the object. The longer the application of the force lasts, the greater distance the object covers, the greater the velocity of the object will be, i.e. the greater the kinetic energy of the object will be.

But according to the special theory of relativity, when f is applied on the object, it finally does not increase the velocity of the object (since acceleration approaches zero), then into what kind of energy the work done by the force f has been transformed?

According to the previous discussion, when v approaches c, the variation of v is very little (Fig.6-3); yet when v approaches c, the variation of m is very obvious (Fig.6-4). That is to say, when v approaches c, though the external force f does not make v vary obviously, it increases the inertial mass of the object. The longer the application of the force lasts, the greater distance the object covers, the greater the mass grows (since m has no upper limit). Therefore, the increase of the energy of the object and the increase of its inertial mass m are related. That is to say, the magnitude of the inertial mass should indicate the magnitude of energy. This is another extremely important deduction of the special theory of relativity.

Three months after the publication of the first treatise of Einstein's special theory of relativity in 1905, he wrote another less than 2000 words paper, treating in particular the relation between the inertial mass and energy. The title of the article is unique, which, if not explained in standard terms of physics, would be: *Does Laziness* (the Inertia of an object) *Depend upon Its Agility* (Energy Content)? Incidentally, in German laziness and inertia are the same word, and so are agility and energy.

His answer is: the inertia of an object is the measurement of its energy. This scientific assertion that smacks the flavour of philosophy, is the famous equation that has been praised as the landmark of the new epoch:

$$E = mc^2$$

wherein E is the energy of the object, m is the mass (inertia) of it, c the velocity of light. It shows that as long as the energy of the object is being increased, its mass will increase proportionally.

In the Newtonian mechanics mass and energy are independent of each other: Between them there is no relation. In relativistic mechanics, energy and mass are only two different aspects of the same mechanical property of the object. To seek internal relations of things that appear entirely different on the surface is the eternal theme of natural sciences.

From the above equation we may notice that even when an object is at rest, its energy differs from zero: $E_r = m_0 c^2$. This energy is called the rest energy. In Newtonian mechanics, only such energies as kinetic energy and potential energy are known, while that form of energy, namely the rest energy has never been mentioned. Rest energy is a certain mode of energy found through the evolvement of the relativistic concept of space and time.

The magnitude of rest energy is fabulously great. The rest energy of an object is generally several hundred million times as great as its chemical energy. If we can only exploit the vital power latent in the object at rest, the source of energy can never be exhausted. With the development of nuclear physics, we have discovered various ways of exploiting the rest energy. For example, the nuclear reactor is one of them. For the time being all the countries are zealously studying the controlled thermonuclear reaction which is also a promising way to release the rest energy.

Now let us recall the journey we have already traversed. From the most scholastic problem whether simultaneity is relative or absolute to the technical problem of controlled thermonuclear reaction, the whole passage closely connecting them is laid upon the foundation of the special theory of relativity. If in this world any truth has woven so much philosophical thought, physical perception and technological application in one simple expression, thereby eloquently revealing the enormous potency of human wisdom, up to the present $E = mc^2$ might serve as the best specimen.

CHAPTER VII FROM THE LEANING TOWER OF PISA TO THE GENERAL THEORY OF RELATIVITY

The Experiment Atop the Leaning Tower of Pisa

Of all the forces in the world the one that was first noticed by man is the gravitational attraction of the earth. The earth attracts all the objects near its surface, pulling them towards itself. Therefore people have been interested in the characteristic of this force since ancient times.

We still have to start from Aristotle. Aristotle in his mechanics once described the characteristic of the gravitational force as follows. He said that when different objects fell down due to the gravitational force, the heavy object falls faster, while the light object fells more slowly. If we let two balls of the same size—one made of

Fig. 7−1 The leaning tower in Pisa

wood, the other of iron—drop at the same time from the same height, then, according to Aristotle, the iron ball will hit the ground first, whereas the wooden ball will hit the ground next. Nevertheless, Aristotle never did the experiment for himself. In his time the cognition of nature was not through the comparison of experiment with theory: speculation was the usual recourse.

Is Aristotle's assertion right or wrong? The solution cannot be completely obtained through speculation. The first man who set to analyse seriously this assertion was again Galileo. He is said to have actually conducted an experiment to test Aristotle's theory.

He carried out the experiment on the top of the Leaning Tower in Pisa[1] (Fig. 7-1). He let objects of different materials fall from the top of the tower to determine the different time required for the fall of each of the objects. He found that all the objects reached the ground simultaneously: none did earlier than the others. That is to say, the falling motion has nothing to do with the concrete properties of the object. Both the wooden ball and the iron ball, if we let them drop at the same time from the top of the tower, will reach the ground at the same time. In this manner Aristotle's theory of gravitational force was disproved by experiments.

Gravitational Force

On this basis Newton advanced the study of the property of the gravitational force. His chief contribution comprises the following two aspects:

The first contribution is conceptional. He tore down Aristotle's partition between "the world above the moon" and "the world below the moon." We have already mentioned this point in the first chapter. Newton maintains that although the falling motion of an object near the surface of the earth differs in appearance from the continuous revolution of the moon, yet both of them result from the same cause— the gravitation of the earth. The reason why gravitation in Newton's theory is called "universal" gravitation is that, by the word "universal", the emphasis is laid on the omnipresence of this force in the universe and the absence of such bounds as defined by Aristotle.

The second is physical. Newton gave a general quantitative description of the gravitational interaction between two objects. If there are two particles of masses

---

(1) According to the research of some of the scientific historians, Galileo never did the renowned Pisa-Tower experiment. In his experiment he used the sloping plane instead of the leaning tower. As a result he found out that it would take different balls made of different materials the same length of time to complete the downward journey on the plane. Yet Pisa Tower has been made "holy" by this misrepresented ancedote in physics, still attracting many pilgrims to pay tribute of reverence to it. Moreover, in Pisa and Florensa some museums still exhibit the wooden balls which were reputed to be the ones Galileo used in his time.

m_1 and m_2 separated by a distance r, then their mutual attraction is

$$F = G \frac{m_1 m_2}{r^2}$$

where G is the gravitational constant, its value being 6.67×10^{-8} dyne. cm·g^{-2}.

Newton's theory of universal gravitation was a great success. According to this theory diverse terrestial and celestial phenomena have been explained, among which the prediction of the existence of Neptune was the most remarkable example. In the early years of the nineteenth century, some perturbations of unknown reason were discovered in the orbit of Uranus. Le Verrier of France and J. Adams of England predicted that the cause might be the gravitational attraction exerted by an undiscovered planet on Uranus. Their mutually independent calculations showed the same result. On Sept. 23, 1846 they forwarded their calculations to the Berlin Observatory, imforming it that a hitherto undiscovered planet should appear at about 5 degrees east of the δ star of Capricornus, its moving speed being 69 arc seconds backward everyday. The Berlin Observatory made the observation accordingly. As a result a new planet of the 8th magnitude was discovered less than 1 degree away from the predicted position. Observation on the following night showed that its motion agreed perfectly with the prediction of Newton's gravitational theory. This success won an unshakable reputation for the theory of gravitation.

Up to now Newton's theory of gravitation remains the basis of precise celestial mechanics. The study of the orbits of artificial satellites and spaceships is still based upon Newton's theory.

In the first years of the 20th century, the gravitational theory seems to be ever-triumphant. Yet there was a very little incident that seemed to defy Newton's authority, i.e. the perihelion advance of Mercury.

The Precession of Mercury's Perihelion

Mercury is the planet nearest to the sun. According to Newton's theory of gravitation, the sun's gravitational force acting on Mercury makes the latter's orbit a closed ellipse. Yet in fact it is not a precise ellipse: with every revolution its major axis also rotates slightly (Fig. 7–2). The rotation of the major axis is called the precession of the perihelion. The rate of the Mercury's precession is $1°33'20''$ per one hundred years. The cause of the precession is the sum total of the gravitational forces of all other planets besides the major gravitational force of the sun acting on Mercury. Yet the former forces are very weak, therefore they can produce only a very slow precession. The astronomers prove in accordance with Newton's gravitational theory that the net result of the advance of Mercury's orbit caused by Earth and the other planets should be $1°32'37''$, instead of $1°33'20''$, per one hundred years. Though the discrepancy is very small, ie only $43''$ per one hundred years, yet it is beyond the negligible amount of observational error.

This $43''$ per one hundred years has given rise to a great deal of debate. Le Verrier, who had successfully predicted the existence of Neptune, turned to this old

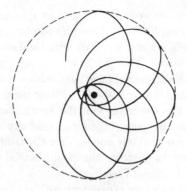

Fig. 7–2 The precession of the elliptical orbit of Mercury

sesame again and predicted another small planet near the sun which would account for the unusual advance. Yet this time Le Verrier failed. No new planet was found where and when as predicted.

In this manner this little 43″/ 100 years has been a puzzle to the celestial mechanics based on Newton's laws. This 43″/100 years, however, is so little that compared with the great successes scored by the Newtonian theoretical system, it may safely be regarded as insignificant.

Yet so far as the scientific problems are concerned, we cannot decide whether a theory is good or bad by how many times it has proved applicable. Hundred thousands times of successes cannot justify the tolerance of a "little" failure.

The problem awaits a solution.

The first satisfactory solution of the Mercury's perihelion advance was not obtained until the establishment of Einstein's general theory of relativity. Yet the study of the general relativity started not with this particular problem. Like the rest of Einstein's scientic work, the general theory of relativity also originated from the speculation on some simple fundamental problems.

Universality of the Ratio: Gravitational Mass Inertial Mass

Though Newton's theory gives a correct quantitative description of the gravitational force, yet in his theory, the most elementary feature of gravitation is not well defined.

Which feature of gravitation, then, is the most important, if we should get to the bottom?

We have seen in several places how Galileo first criticized, from various angles, the fundamental mistakes in the Aristotelean system, and how Newton, expounding his predecessor's view, developed a complete system of classical physics. In the previous discussions we have also learned that though Galileo only gave some essential concepts which were yet to be woven into a perfect system, yet some notions laid down by him are not only applicable to Newton's mechanics but also correct in the theory of relativity. Such is the case not only with Galileo's principle of relativity, but also with his law of inertia. Though Newton's concepts of absolute time and space, and subsequently his mechanics, have been modified in the theory of relativity, yet these concepts of Galileo's are still valid without the slightest modification.

The case is more or less the same with the development of the gravitational theory. As we shall see, in the general theory of relativity, the formula of gravitation given by Newton is not perfectly correct. Yet the truth discovered by Galileo atop the Pisa leaning tower has become the most essential thing that serves as the starting point of the general theory of relativity.

What does the Pisa leaning tower experiment say?

By comparing Newton's mechanic equation with his law of gravitation, we can describe the motion of the freely falling object by the following equation:

$$m_i a = m_g \frac{GM}{r^2}$$

where m_i and m_g represent the object's inertial mass (inversly proportional to acceleration) and the gravitational mass (directly proportional to gravitation) respectively, M is the gravitational mass of the earth, and r is the distance of the object to the centre of the earth. The above equation can also be written as

$$a = \frac{m_g}{m_i} \left(\frac{GM}{r^2} \right)$$

Pisa experiment shows, no matter what object is chosen, the acceleration of the object produced by the gravitational force is the same. From the above equation, it can be readily seen that the value of m_g/m_i of each and all objects should also be the same. In other words

$$\frac{\text{gravitational mass}}{\text{inertial mass}}$$

is an universal constant, independent of the property of the given object.

In physics, from the discovery of a universal constant a whole set of theory is often to be developed. From the universal c of the light velocity the special theory of relativity has been derived. From Planck constant h the quantum theory has been deducted. The universl constant of m_g/m_i is the key to the solution of the gravitational problems.

Thus wrote Einstein: "...In the gravitational field all objects have the same acceleration. This law can also be understood as the law of equivalence between the inertial mass and the gravitational mass. Even at that time it impressed me with all its important implication. I greatly wondered at its existence, and guessed it must contain a key to the deepening of our understanding of inertia and gravitation".

The Nature of Gravitation is the "Null" of Gravitation

But how did Einstein make use of the key—"m_g/m_i being a universal constant"?

Like Galileo, Einstein devised an ideal experiment to analyze the problem, the only difference between them being that while the former used a slope, the latter used an elevator cabin. Einstein's ideal elevator cabin is installed with various sorts of experimental instruments. In the cabin an experimental physicist can leisurely carry on with all kinds of measurement.

When the elevator is at rest relative to the earth, the experimenter can see that all the objects therein are subject to a force. If there is no other force to balance it, it will pull all these objects onto the floor of the cabin. Moreover, all the objects in the course of falling possess the same acceleration. According to these phenomena the experimenter may readily conclude: his elevator is subject to an external attraction.

Now let the elevator be in a free-falling motion. Then the experimenter will discover that the objects in the cabin are no longer subject to the force as they used to and all objects have lost their usual acceleration. In other words they have entered what we usually call the state of "weightlessness". The objects in the cabin no longer show any sign of being subject to the gravitational force. Whether an apple or a feather may float freely in the air instead of falling downward. The experimenter may either walk the floor or skim the ceiling; in either case he enjoys perfect ease, and to him the skill of an acrobat is superfluous. This is to say, the experimenter in observing the mechanic phenomenon of any object cannot detect the slightest trace of the gravitational force.

Following this Einstein made a further deduction that the experimenter in the cabin not only fails to detect any trace of gravitation through observing all the mechanic phenomena but also, as he affirms, in all other kinds of physical experiments. That is to say, in the frame of the elevator, gravitation has been completely nullified. The elevator experimenter cannot decide by observing physical phenomena within the cabin whether there is an earth outside acting as the origin of the gravitational force, neither can he measure whether his elevator has an acceleration or not, just as in Salviati's ship the observer cannot make out if his ship is in motion or at rest.

In brief, we can find in any local region (as regards the implication of the word local, more will be said later) a frame of reference (Einstein's elevator) wherein all the effects of the gravitational force may completely be eliminated, which fact is the most important property of gravitation. In physics no other force has this nature. For example, neither the macroscopic electromagnetic force nor the strong interaction and weak interaction in the realm of nuclei and particles can be completely eliminated through the choice of anproper frame of reference.

The property of gravitation consists in that it can be eliminated locally in a certain frame of reference (Einstein's elevator cabin). This is the essential property of gravitation that Einstein had abstracted from the facts provided by the Pisa Leaning Tower experiments. This is commonly known as the principle of equivalence.

Fig. 7-3 Einstein's ideal elevator cabin

Local Inertial System

The principle of equivalence ensures that at any moment and at any point in the space there must exist an Einstein's elevator, wherein all phenomena appear as if in the universe no such a thing as gravitation existed. In such an elevator what is moving will be moving forever, ie the law of inertia here is valid. According to the definition, a frame of reference wherein the law of inertia is valid is an inertial frame of reference. Hence Einstein's elevator should be an inertial frame of reference.

Here doubt may arise in your mind, since we usually regard the uniformly moving ship of Salviati's as the inertial frame of reference. But relative to the earth, which is the same as relative to Salviati's ship, Einstein's elevator is not in uniform motion: it has an acceleration (that of the free-falling body). Do they contradict each other?

Yes, they do. Before the development of the general theory of relativity, Salviati's ship has been assumed to be an inertial frame of reference. Yet, strictly speaking, it is a wrong assumption. Since in the Salviati's ship the experimenter will notice that the water drops are making an accelerated downward motion (look, the ship is completely sealed, wherein the experimenter cannot tell whether there is anything outside). That is to say that the water drops do not satisfy the law of inertia that a moving object will forever move uniformly, hence it cannot be regarded as a real inertial frame of reference (at best it can only be regarded as an approximate inertial frame of reference). On the contrary, in Einstein's elevator cabin, the law that what is moving will move uniformly forever can find a full expression.

Now let us turn to the word "local". When we say that the acceleration of all objects caused by gravitation is the same, we mean these objects are at the same point in the space. With respect to various points, gravitational acceleration differs. For example in Fig. 7-4 the gravitational acceleration differs at different points of the earth. Therefore, a freely falling cabin can only completely eliminate the effect of gravitation (e.g. the gravitational acceleration) in a small region round a given point but cannot completely eliminate it in a large region. For example, in Fig. 7-4 the cabin can eliminate the gravitational force at point A, but will produce no effect at point B.

Therefore, if we regard the aforesaid Einstein's cabin as the inertial frame of reference in the strict sense, then this frame of reference is only applicable to a local region. The elevator at point A is only the frame of reference at point A. As to the

Fig. 7-4 The gravitational acceleration is different at different places.

frame of reference at point B, only the employment of the freely falling elevator at point B can serve its purpose.

What is Gravitation?

Now we may try to answer the abstruse question —What is gravitation?

Here let us once more recall that famous speech of Salviati's in which he says "let the ship sail at any speed, as long as the motion is uniform..." This shows that Salviati's ship is only allowed to make uniform motion. In other words, previous to the development of the general theory of relativity, people maintained that between various frames of reference (such as Salviati's ship) only relative uniform unaccelerated motion is allowed. Newtonian mechanics, including the Newtonian theory of gravitation, is all constructed on this basis.

But the general theory of relativity shows that the inertial system in the strict sense can only be some local inertial system (such as Einstein's cabin). Now various local inertial systems at various points may have relative accelerated motion between them. For instance, in Fig. 7–3, between the cabin at point A and that at point B there is an accelerated motion.

Then what is gravitation? The effect of gravitation is that it decides the relations between various local inertial systems. In any of these systems we cannot perceive the effect of gravitation. Only in the interrelations between these local inertial systems can we perceive it.

In other parts of the physics our working plan is always like this: by adopting a certain frame of reference to measure the relevant physical quantities; secondly, through experiment to sum up the rules governing them; finally, to find out the essential equation of these quantities. In this procedure the geometrical properties of space-time (i.e. the adopted frame of reference) are not to be affected by the physical process under discussion. Therefore, in these problems the basic equation will simply be the relations between the relevant physical quantities, i.e.

$$\text{some physical quantities} = \text{other physical quantities}$$

But, as far as the problem of gravitation is concerned, the gravitation will affect the motion of various objects on the one hand, and will affect the relations between the local inertial systems on the other. Therefore we now find ourselves unable to define the geometrical properties of space-time beforehand, their properties being the vary things to be decided afterwards. Therefore, the basic equation of gravitation cannot but involve the geometrical properties of spacetime. It should reflect gravitation itself as well as the interaction between gravitation and matter, i.e. it should be an equation of the following form:

$$\text{geometrical quantities} = \text{physical quantities of matter}$$

Einstein's Gravitational Field Equation

Einstein had spent some seven or eight years seeking this fundamental equation of gravitation, during which time he had been confronted with failures repeatedly. Towards the end of 1915, he found at last what he considered as the satisfactory equation of gravitational field. At that time he wrote a letter to A. Sommerfeld, saying: "Last month is one of the most thrilling the most intense periods of my life. What makes me so happy is, not only Newton's theory turns out to be the first approximation, but the perihelion precession of Mercury (43" per one hundred years) has been obtained as the second approximation."

From pisa Leaning Tower up to 43" per 100 years, the relations between these and other evidences were finally obtained.

The entire course of Einstein's labour for the acquisition of the gravitational field equation is an epic worthy of our study. The methodology characteristic of Einstein's work is highly enlightening. Yet in this little book we cannot dwell on it in detail, since a thorough discussion of it will inevitably involve a lot of mathematical apparatus. Here let us just write down its final result:

$$R_{\mu\nu} = -8\pi G(T_{\mu\nu} - \frac{1}{2}g_{\mu\nu}T_{\lambda}^{\lambda})$$

where $g_{\mu\nu}$ is called the metric tensor, $R_{\mu\nu}$ the Ricci tensor—quantities employed to describe the geometrical properties of space-time and $T_{\mu\nu}$ the energy-momentum tensor, a physical quantity employed to describe the physical property.

All in all, in Einstein's general theory of relativity, space, time and the motion of matter are interacting upon each other. Here the theory has not only freed itself from the absolute space-time independent of the motion of matter as Newton understood it, but has also excelled the elementary relativity suggested by Salviati's ship. Einstein once said "Space-time may not be considered as something that can be separated from the real objects of physical entity and exist of itself. Not that the physical objects are **in the midst of space,** but that these objects **claim the spatial extensions.** Therefore the conception of an empty space has lost its significance."

This is his scientific as well as philosophic conclusion.

CHAPTER VIII   FROM NEWTON TO POST-NEWTON

Post-Newtonian Corrections

Though the basic conception of Einstein's general theory of relativity differs from that of Newton's theory of gravitation, yet, with the domain where Newton's theory is applicable, the results of both theories should be in agreement. As we have already mentioned, Newton's theory of gravitation is a fairly good theory, it correctly explains numerous phenomena.

When we say the realm wherein Newton's theory is applicable, we only mean, to be exact, a weak gravitational field.

Yet how should we delimit the intensity of gravitational field? Roughly speaking, if the velocity attained by any object arising out of the action of the gravitational field is far below the velocity of light, the field is a weak one. On the contrary, if the velocity is comparable to that of light, the field is a strong one.

The orbital velocity of the earth is only 20 km/s, by far below that of light (300000 km/s), and therefore the gravitational field of the sun is a weak one. Generally speaking, in the gravitational field of an object of mass M, the orbital velocity is approximately

$$v = \sqrt{\frac{GM}{R}}$$

where G is the gravitational constant and R is the spatial dimension of object M. From this equation it can be readily seen that the condition of a weak field is $\sqrt{\frac{GM}{R}} \ll c$, or

$$\frac{GM}{c^2 R} \ll 1$$

whereas the condition of a strong field is

$$\frac{GM}{c^2 R} \approx 1$$

The following table shows the $\frac{GM}{c^2 R}$ values for some common objects:

object	proton	man	the earth	the sun	the Milky-Way Galaxy
$\frac{GM}{c^2 R}$	10^{-40}	10^{-25}	$10^{-8.9}$	$10^{-5.4}$	10^{-6}

All these values are far smaller than 1. This is the reason why Newton's gravitational theory is applicable to a great number of problems.

As for Einstein's gravitational equation, in the case $\frac{GM}{c^2R} \ll 1$, it should assume the form of Newton's law of gravitation. For instance, the planets that move in the gravitational field of the sun are subject to the gravitational force of the latter. This force can be described by the formula $F = G\frac{m_1 m_2}{r^2}$. We can also describe the field by the potential energy between the sun and the planet. According to Newton's theory, the potential energy is

$$U = -\frac{GmM_\odot}{r}$$

where m is the mass of the planet, $M_\odot$ is the mass of the sun, and r is the distance between them.

According to the general theory of relativity, the potential energy resulting from the gravitational interaction should take the revised form:

$$U = -\frac{GmM_\odot}{r} - \frac{3}{2}\frac{v^2}{c^2}\frac{GmM_\odot}{r} + \ldots\ldots$$

in which the first term is the same as shown by Newton's theory, while the second term is the correction made by the general theory of relativity and is very small compared with the first term, since $\frac{v^2}{c^2} \approx \frac{GM}{c^2R} \approx 10^{-6}$ (see the $\frac{GM}{c^2R}$ value for the sun in the above table). If the second term is neglected, the formula will reduce to Newton's law of gravitation.

In the above formula, the first term is called the Newtonian term, the second, third, etc are called the post-Newtonian terms. In the case of $\frac{GM}{c^2R} \ll 1$, the second term is only a small correction that the general theory of relativity has given to the Newtonian theory. This correction is called the post-Newtonian correction.

The Precession of Planetary Perihelia

Though the post-Newtonian correction is very small, it sometimes plays a key role. The anomalous precession of Mercury's perihelion is explained by the post-Newtonian term. If there only existed the Newtonian term, the anomalous precession of Mercury's perihelion would not take place.

Not only the anomalous precession of perihelion has been observed for Mercury, but also quantitative results of measurement in the case of some other planets have been obtained. The following table gives the observational results compared with the computed post-Newtonian corrections. Here we see that theory and observation are in fairly good accord.

planets	observation	theory
Mercury	43″.11 ± 0″.45/century	43″.03/century
Venus	8″.4 ± 4″.8/century	8″.6/century
Earth	5″.0 ± 1″.2/century	3″.8/century
Icarus (asteroid)	9″.8 ± 0″.8/century	10″.3/century

The Precession of the Axis of Rotation

In Newton's mechanics, the rotation of the planet does not participate in the gravitational interaction. That is to say, the gravitational force the sun exerts on the planet is related only with the mass of the planet, and has nothing to do with the speed of rotation of the planet. Newton's equation of gravitation contains the mass, and not the speed of rotation, of the object as a factor.

Yet the general theory of relativity is different. In some of the post-Newtonian correction terms, not only the mass, but also the physical quantity of rotation of the object are included: the speed of rotation contributes to gravitational effect. The

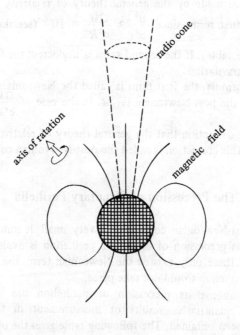

Fig. 8–1 The magnetic axis and the axis of rotation of the pulsar do not lie in the same direction. The cone-like radio beam lies in the direction of its magnetic axis. Therefore, in the course of its rotation, anytime the radio cone points to the earth, we can receive a radio pulse

gravitational interaction between two particles without rotation differs from that between the particles when rotating.

This new feature may give rise to the precession of the axis of rotation. That is to say, the direction of the axis of rotation of the planet in the course of revolution around a sun may change slowly. As regards the planets in the solar system, this post-Newtonian effect is generally so small that it can hardly be measured. Moreover, there are other factors that may produce the change of the axis of rotation. All these help to drown out the post-Newtonian contribution.

On the other hand, the observation in recent years on the pulsar PSR1913+16 has produced qualitative evidence concerning the precession of its axis of rotation. PSR1913+16 consists in fact of two compact stars (as to compact stars, we shall discuss them in the next chapter). One of them is a fast rotating radio pulsar. The emission of a pulsar is a concentrated cone-like beam (see Fig. 8–1). With each rotation of the pulsar the cone sweeps across the earth once and we shall as a result receive a radio pulse.

After PSR1913+16 was discovered towards the end of 1974, observation during subsequent years indicates that its radio pulse (or pulse contour) has shown some change (see Fig.8–2). This may be the result of the advance of the axis of rotation.

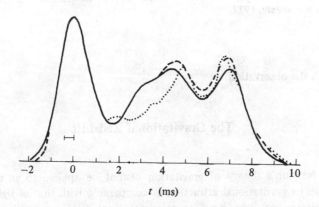

Fig. 8–2 The changes of PSR1913+16 pulse contour. The dotted line is the result of measurement in July, 1977; the broken line and the real line represent the results of measurement in June and October of 1978 respectively

Since the cross-section of the radio-cone has roughly the form shown in the following picture, when the axis of rotation precesses, the area swept by the cone will be different. The picture marks out the possible sweeping lines in July 1977 and October 1978. Therefore from the change of pulse contour we can estimate the magnitude of the precession in the axis of rotation. According to the calculation of post-Newtonian correction, the rate of precession should be 1°/year. The value

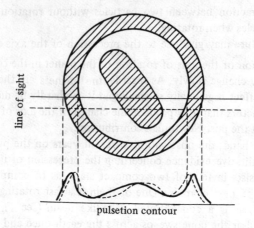

Fig. 8–3 An explanation of the pulsation contour: the shadowed areas represent the cross-section of the radio-cone of the pulsar, the horizontal line represents the track of the observer's eyesight when the pulsar is rotating. When the axis of rotation precesses, areas that the eyesight penetrates change accordingly, hence the changes of the pulsation contour seen by the observer. The real line represents the observed line of October, 1978; the broken line represents the observed line of July, 1977

agrees with the observation.

The Gravitational Redshift

Since Newton's theory of gravitation cannot be applied to an object whose velocity due to gravitational attraction is comparable with that of light, it follows that the motion of light itself in a gravitational field cannot in principle be described by Newton's theory of gravitation. The interaction between light and gravitational field pertains essentially to the post-Newtonian terms. In the last few sections of this chapter we shall discuss several new phenomena concerning the propagation of light in a gravitational field.

The first to be discussed is gravitational redshift.

By this effect we mean that when light propagates in the gravitational field, its frequency or its wavelength will change. When the light emitted by a hydrogen atom on the surface of the sun reaches us, we shall notice that its frequency is lower than that of light emitted by a hydrogen atom in the laboratory, or that it has redshifted (in the visible light the frequency of red light is the lowest, therefore the decrease of frequency is usually called redshift, whereas the contrary is called blueshift). This is because the gravitational field on the surface of the sun is stronger than that on the earth (or the $\frac{GM}{c^2 R}$ value of the sun is greater). If someone can stand on the surface of the sun to receive light emitted from the earth, he will discover that the frequencies of the light become higher, or that they have blueshifted.

In short, when light goes from a strong gravitational field ($\frac{GM}{c^2 R}$ being great) to a weak gravitational field ($\frac{GM}{c^2 R}$ being small), the frequency becomes lower. The other way round it becomes higher.

Since 1960 quantitative tests have been performed on the gravitational redshift theory. R. Pound and others placed a γ-ray source of ^{57}Co at the base of a 22.6 m tower, and a receiver of ^{57}Fe at the top of the tower. This type of Mössbauer experimental device[1] can achieve a frequency stability of the order of 10^{-12}. When the γ-ray emitted by ^{57}Co reaches the top, a slight redshift will take place. Measurement and theoretical prediction agree beautifully; the ratio between the experimental and the theoretical value is 0.997±0.008.

The Deflection of Light

In the gravitational field no object can move along a straight line, since the effect of gravitation will make its track curve towards the source of gravitation. By the principle of equivalence, we can assert that the same thing will happen to the propagation of light in the gravitational field. This is because, if the motion of light differs in manner from that of other objects, it will be impossible for us to find an Einstein's elevator cabin wherein the effect of gravitation can be eliminated both in the motion of ordinary objects and in that of light. Therefore, from the prerequisite of the existence of a local inertial system in which the gravitation can be eliminated it follows as a corollary that the passage of light in the gravitational field should also be deflected.

A beam of starlight that passes through the gravitational field near the surface of the sun deviates only by an angle of 1."75. If the sun is not between the star and the earth, the starlight will come to us along a straight line. But when the sun is in between, the passage of the starlight will be deflected and we find the apparent position of the star shifted to where the broken line indicates, as shown in Fig.8—4.

In 1919 a surveying expedition led by Sir Arthur Eddington for the first time verified the predicted effect. On May 29 of that year from the island of Principe, West Africa they photographed the star-field in the neighbourhood of the sun during

---

(1) When the nucleus emits γ-ray, due to the recoil of nuclei, the energy of the γ-ray is somewhat smaller than the transition energy between two different levels. Therefore the emitted γ-ray cannot be absorbed again by the corresponding resonance levels. In order to overcome the effect of this nuclear recoil, Mössbauer had the emitting nuclei inlaid in a large block of crystal. Since the mass receiving the recoil has been greatly increased, the energy decrease of γ-ray due to the recoil absorption is reduced, thereby making possible the above-said resonance.

Fig. 8–4 When the sun appears between the star and the earth, the passage of the starlight will be deflected

its total eclipse, and then compared it with the plate taken when the sun was not in this region of the firmament. From this comparison they obtained an estimate of the light deflection which was in fair agreement with the theoretical prediction.

Ever since 1919, whenever a chance turned up for observing a total eclipse of the sun, astronomers of many countries would attempt this light deflection experiment. The following table shows the chief results of these observations:

date	place	observational data
May 29, 1919	Brazil	$1''.98 \pm 0.16$
May 29, 1919	Principe	$1''.61 \pm 0.40$
Sept. 21, 1922	Australia	$1''.72 \pm 0.15$
May 9, 1929	Sumatra	$2''.24 \pm 0.10$
June 19, 1936	Soviet Union	$2''.73 \pm 0.31$
June 19, 1936	Japan	$1''.28 \pm 2.13$
May 20, 1947	Brazil	$2''.01 \pm 0.27$
Feb. 25, 1952	Sudan	$1''.70 \pm 0.01$
June 30, 1973	Mauritania	$1''.60 \pm 0.18$

In recent years the spatial resolution applied by radio-astronomy has been so much improved that its differentiating rate excels that attained in optical astronomy. For this reason the deflection of light can now be ascertained with higher precision. Fortunately in March and April every year the sun passes by the radio-source

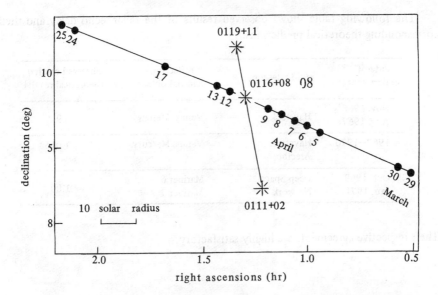

Fig. 8–5 The sketch of the radio-sources 0116+08, 0111+02 and 0119+11 and the location of the sun (when the sun passes near the radio-source 0116+08, according to the changes of the relative positions of the three radio-sources, the data of the light deflection in the gravitational field can be obtained)

0116+08 once (see Fig. 8–5). Since the three radio sources 0116+08, 0119+11 and 0111+02 almost lie in a straight line, their relative positions will be changed when the sun passes by 0116+08. By this method, the obtained datum of light deflection is $1''.775 \pm 0''.019$.

The Time Delay of Radar Echo

In the year 1964, I. Shapiro and others proclaimed a new testable effect of the propagation of light in the gravitational field.

Shapiro used a radar to emit a beam of electromagnetic waves which bounced off another planet to be received again by the same radar. One can measure the times for a round trip in two different cases: in the first the passage of the electromagnetic waves is far away from the sun so that the influence of the latter is negligible, and in the second case the round trip of the electromagnetic waves cuts through solar neighbourhood where they are subject to the influence of a strong gravitational field. In the latter case the arrival of the echo is somewhat later than that in the former case. This lengthening of the propagation time caused by the gravitational field of the sun is called the delay of the radar echo. For instance the delay of the radar echo between the earth and Mercury can be as long as 240 miliseconds. In order to avoid the effect of complex factors of the surface of the planet, artificial celestial object has also been used to serve as the reflecting target of the radar signal.

The following table shows observed results of the radar echo delay and their corresponding theoretical predictions:

date of experiment	radio telescope	reflecting celestial object	observed results/ theoretical results
Nov. 1966– Aug. 1967	Haystack	Venus, Mercury	0.9
1967–1970	Haystack Arecibo	Venus, Mercury	1.015
Oct. 1969– Jan, 1971	Deep Space Network	Mariner 6 Mariner 7	1.00

Their respective agreements are highly satisfactory.

72

CHAPTER IX FROM CLASSICAL GRAVITATIONAL COLLAPSE TO THE BLACK HOLE

Something More About the Strong Field Requirement

In the previous chapter we have pointed out that the strong field requirement is $\frac{GM}{c^2R} \approx 1$.

Now let us look at the problem from another angle. If the gravitational field produced by a system of mass M is strong, its spatial dimension should be $R \approx \frac{GM}{c^2}$. In other words, if M is the origin of a strong field, then M should be compressed into a volume as small as $R \approx \frac{GM}{c^2}$. The following table shows the $\frac{GM}{c^2}$ value of some of the objects.

system	proton	man	earth	sun	the Milky Way
$\frac{GM}{c^2}$ (cm)	10^{-52}	10^{-23}	10^{-1}	10^{5}	10^{16}

According to our laboratory experiences, it is out of the question to realize the required compression shown in the above table. The most powerful compressor of today cannot even reduce the volume of water by more than 10 per cent. Any attempt at compressing the gigantic sun into a ball only a few km in diameter sounds but like a fairy tale.

Yet, in nature, is there any powerful compressor that can compress a weak field system into a strong one? The above experience suggests a negative answer. As a matter of fact, concerning whether there really exist in nature strong-field systems, some physicists are still rather doubtful. If strong-field objects do not exist at all, the general theory of relativity, no matter how good it is, will look just as miserable as a hero who can find nowhere to display his valour and might.

Sometimes, however physicists place more trust in general laws that have stood the test than in everyday experiences. Common experiences sometimes are unhelpful when uncommon problems await our judgment. As regards the problems of compression, the results obtained from physical laws directly challenge our everyday impression: in nature not only exist extraordinarily powerful compressors, but also most of the celestial objects cannot escape the destiny of being compressed. The compressors are none other than the gravitational fields of the doomed celestial objects themselves.

Gravitational Collapse

The problem to be discussed originates from the analysis of stellar equilibrium.

The properties of a star are chiefly decided by two main forces: the gravitational force of the star itself and the pressure of the matter within the star. If the pressure is greater than the gravitation, the star will expand; if the gravitation is greater than the pressure, the star will contract; if these two forces are equal, the star will reach a state of equilibrium.

As early as 1930, E. Milne analyzed a star made of classical ideal gas without energy sources. He found that in this specific condition pressure can never match gravitation. System of any mass under the action of its own gravitation will collapse indefinitely until its spatial dimension comes to zero and the density of matter increases to infinity.

After this S. Chandrasekhar and L. Landau independently pointed out that Milne's analysis was not perfect, since in the state of high density, the property of matter can no longer be described by the model of a classical ideal gas. In that state of matter the principle of exclusion in quantum mechanics should be taken into consideration. This principle of exclusion can lead to an enormous force to resist the collapse. The force is usually called the degeneracy pressure. More exactly, degeneracy pressure in the high density condition can be grouped into two classes. The first class is electron degeneracy pressure which plays the major role when the density of matter is within the range of $10^4 - 10^8 \text{g/cm}^3$; another class is neutron degeneracy which plays the major role when the density of matter is within $10^{12} - 10^{15} \text{g/cm}^3$. Computation shows that when degeneracy pressure has been taken into consideration, the situation seems to improve: when the mass of the celestial object is within a certain range, Milne's indefinite collapse will not appear. Chandrasekhar's calculation indicates that when a certain spatial dimension is reached, the degeneracy pressure will balance the self-gravitational force. The compact star thus balanced is called the degenerate dwarf star, the white dwarf being an example. Sirius β is a white dwarf. Yet electron degeneracy cannot absolutely ward off the threat of collapse. This is true for those celestial objects whose masses exceed 1.5 solar mass, and they can no longer form stable dwarf stars. The fate awaiting them is still indefinite collapse. Thus Chandarsekhar described the situation: "Our conclusion is , not until we can answer the following basic question,can the analysis of stellar structure be further advanced. The problem is: if a closed body that contains electrons and nuclei (so that the whole body is neutral electrically) is indefinitely compressed, what will happen next?"

All the above discussions are based upon Newton's theory of gravitation.

Towards the end of the thirties, Oppenheimer in his analysis of the problem made use of the general theory of relativity, but the result proved to be unaltered. Though he and others proved the existence within a certain range of mass of stabilized neutron star (i.e. star resulting from the equilibrium between neutron degeneracy pressure and gravitational force), he added: "When all the thermonuclear energy sources are exhausted, a sufficiently massive star may indefinitely collapse."

Once the notion of indefinite collapse is released from the "Physics' Box" like disasters set free from Pandora's Box it can never be recalled. In short, there are two conclusions regarding the fate of the stars:

1. Gravitational collapse occurs, hence the formation of a great number of compact objects.

2. The compact objects can be grouped into two classes: some are formed as the result of finite collapse, such as the white dwarfs and the neutron stars, others are formed as the result of indefinite collapse.

As for the first conclusion, it can be obtained from both Newton's theory of gravitation and the general theory of relativity. The second conclusion can only be deduced from the general theory of relativity, since Newton's theory of gravitation is not applicable to the case of strong field. Now let us discuss the observed evidences supporting the first conclusion.

Where are the Strong-field Objects?

In the year 1934, W. Baade and F. Zwicky published a short article in which they put forward some conjectures concerning the search of these strange celestial objects. The article is so short, in content so comprehensive, and in prediction so bold and accurate that the history of physics and astronomy can boast of very few examples of its like. Here we would rather cite the original text than reiterate its viewpoints.

SUPERNOVA AND COSMIC RAY

Supernovae flare up in every stellar system (nebula) once in several centuries. The lifetime of a supernova is about twenty days and its absolute brightness at maximum may be as high as $M_v=-14^m$. The visible radiation L_v of a supernova is about 10^8 times the radiation of our sun. That is $L_v=3.78 \times 10^{41}$ erg /s. (Calculations indicate that the total radiation, visible and invisible, is of the order $L_r=10^7 L_v=3.78 \times 10^{48}$ erg /sec. The supernova therefore emits during its life a total energy $E_r \geqslant 10^5 L_r=3.78 \times 10^{53}$ ergs. If supernovae initially are quite ordinary stars of mass $M<10^{34}$ g, E_r/c^2 is of the same order as M itself. In the supernova processes mass in bulk is annihilated. In addition the hypothesis suggests itself that cosmic rays are produced by supernova. Assuming that in every nebula one supernova occurs every thousand years, the intensity of the cosmic ray to be observed on the earth should be of the order $\sigma=2 \times 10^{-3}$ erg/cm$^2 \cdot$s. [1] The observational values are about $\sigma=3 \times 10^{-3}$ erg/cm$^2 \cdot$s. With all reserve we advance the view that supernova represent the transitions from ordinary stars into neutron stars, which in their final stages consist of extremely closely packed neutrons.

W. Baade and F. Zwicky, Phys. Rev. 45, (1934), 138.

Observation and research in the following thirty years have proved the correctness of Baade and Zwicky's paper. The key evidence has come from the study of the Crab Nebula.

The Crab Nebula is an expanding gaseous nebula in the Milky Way. Its luminosity is very great, roughly equal to that of 100 suns together. From where does the nebula obtain its energy? The problem has attracted the attention of many astronomers.

As early as 1928, the idea was put forward that the Crab Nebula and the 1054 supernova (see Chapter III) are related. Later on it was discovered that the nebula is still expanding. From the expanding rate it can be calculated that the expansion from its very beginning up to now has taken some 800 years. This figure is near to the duration from 1054 to the present, thus convincingly supporting the idea that these two things are related. But how are they related?

After this people set to study a star in the midst of the Crab Nebula. This star is very strange. Its luminosity is very great, about 100 times that of the sun. Yet it shows no spectral line in its spectrum. Its spectrum is entirely different from that of a common star.

Up to then it seemed that in the study of the Crab Nebula problems abound whereas solutions are few. What has the 1054 supernova explosion left behind? From where does the nebula radiation obtain its energy? To what class does the star in the midst of it belong? All these problems await solution. Yet it seems that the more problems we are confronted with and the sharper these problems are, the nearer we are to solution.

The key step towards the solution is to have the variation of luminosity measured out. By high-speed light measurement we discover that the luminosity of the star in the Crab Nebula varies continually: the variation is very regular, its exceedingly stable period T being

$$T = 0.03310615370 \text{ s}$$

Up to the present this is the shortest period ever found in the celestial phenomena.

Pulsar is a Kind of Compact Object

From the stability of the period we can conclude that the pulsation is produced by the rotation of the object. The shortness of the period indicates the smallness of the spatial dimension of the object. On the other hand, its great luminosity shows that its mass cannot be too small. Is not this object of great mass and small size a compact object born out of gravitational collapse?

This realization greatly facilitates the solution of many problems concerning the Crab Nebula. (1) The star descended from a common star in the course of the supernova explosion in 1054. The period of rotation of a common star is generally a month. Due to the constancy of angular momentum, in the process of collapse the angular velocity increased continually. Therefore after a compact star was formed, its rotation period was reduced to several milliseconds. (2) Accurate measurement reveals that the pulse period tends to lengthen, though almost imperceptibly—a tendency that betrays the slowing down of the rotation of the compact star and the gradual decrease of the rotational energy. The loss in rotational energy is precisely equal to the energy radiated by the central star and the nebula.

These results satisfactorily support the view of Baade and Zwicky: supernova is the phenomenon that appears when a common star collapses into a compact star.

Though the star at the centre of the Crab Nebula is not the first pulsar that has been discovered, the important conclusion that a pulsar is a kind of neutron star has been derived from the study of the Crab Nebula. It is very interesting to note that, though during the last few decades a number of astronomers have observed the Crab Nebula, its property of pulsating luminosity remained unknown till recently. This is not strange, since the effect of visual retention of human eyes makes them unable to detect luminosity variation period shorter than 60 milliseconds. The 33-millisecond period of the pulsar of the Crab Nebula is just below the limit. If the period of the Crab Nebula pulsar were somewhat longer, the discovery story of the compact star might have begun long before. So Nature designed everything, as if she is bent to test the human intelligence.

Actually the discovery of the neutron star is the fruit of the versatile intelligence of mankind. So far as physical theory is concerned, all the theories from classical physics to the theory of relativity have been resorted to. As for technology, it involves astrometry spectral analysis and time keeping in addition to the faithful and exhaustive records of Chinese astronomers of the 11th century.

Today more than three hundred pulsars have been recorded in the Milky Way which presumably contains more than 10^9 compact stars of this kind.

Now we have verified the first theoretical prediction concerning strong gravitational field: That there exist a great number of compact stars resulting from gravitational collapse.

Now let us turn to the second problem: Do there exist two classes of compact stars created as the result of finite collapse and infinite collapse separately?

Before we exhibit the observed evidences, let's discuss more carefully the theoretical prediction concerning the finite and infinite collapse.

The Structure of Neutron Stars

The finite collapse may result in the formation of the white dwarfs, the neutron stars, abnormal neutron stars, quark stars, etc. The reason why so many names are designated to these compact stars is because we do not know very clearly about them, hence the conclusions do not agree completely. Yet they have many things in common. Here let us discuss the neutron star, since it is typical of them.

The collapse of a star whose mass is greater than 1.4 solar mass generates very great pressure. Under this pressure, electrons in the atoms are almost completely captured by protons in the nuclei, releasing neutrinoes and turning into neutrons. Therefore the whole star consists almost entirely of neutrons. Its density is a million million to hundred million million times that of water (i.e. $10^{12}-10^{14}$ g/cm^3). The diametre of a neutron star whose mass is about the same as the sun measures only about several km.

Since all the stars have rotation and magnetic field, when any of them has collapsed into a neutron star, its rotation will be speed up (due to the conservation of angular momentum). Its magnetic field will be intensified, since the magnetic field that used to be distributed over a large region within the star will be compressed into a small volume after its collapse. For a star similar to the sun, its magnetic field will be increased over ten thousand million times after it has collapsed into a neutron star.

So a neutron star is usually a fast rotating star with a very strong magnetic field. Generally speaking, the direction of magnetic axis does not agree with the axis of rotation, just as the axis of rotation of the earth does not agree with its magnetic axis. In the neighbourhood of the magnetic poles of the neutron star, the magnetic field is exceedingly strong. Electrons moving in this strong magnetic field will emit intense radio waves. The radio emission mainly lies in the direction of the magnetic axis. During the interval when the magnetic axis of the neutron star points more or less towards the earth, the latter will receive the radio waves. With each rotation of the neutron star, we shall receive one signal. This is how the pulsating radio waves are formed (Fig.9−1).

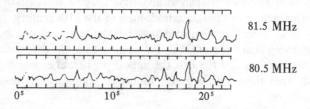

Fig.9−1 The pulsating signals of CP1919, the first discovered pulsar

These are the chief characteristics of objects formed as the result of the finite collapse.

The Black Hole

The end point of the in finite collapse is the black hole.

As early as 1795, the French astronomer, mathematician and physicist Pierre Laplace pointed out: light cannot escape from the surface of a star of sufficiently great mass. According to Newton's gravitational theory, for every object there is a definite escape velocity. The escape velocity of the earth is often called the second cosmic velocity, i.e. 11 km/s approximately. For a celestial object of great mass and small dimensions, the escape velocity may be greater than that of light. Therefore, viewed from outside, the celestial object does not emit light. It may be called a black hole in the Newtonian theory. Yet, as we already know, Newton's gravitational theory in principle cannot tackle the problem of light. Therefore the conclusion cannot be taken for granted.

In the general theory of relativity there is also the process of infinite gravitational collapse. Let us imagine a man stands on the surface of a collapsing star, holding a powerful lamp in his hand. Before the collapse the gravitational field is weak, the lamp in his hand is throwing out its light in every direction. The beams of light are virtually travelling along straight lines (Fig. 9–2). But as soon as the star begins to collapse, its mass starts to be concentrated in a ever shrinking volume. As the size of the star diminishes, its surface gravity will become greater and greater, causing the track of light to curve. In the beginning, only the beams in the horizontal directions show any obvious curvature and do not leave the star but return to its surface. As the collapse goes on, more and more beams of the lamp will curve back to the star. Finally of all the beams of light none can escape the surface of the globe. At this point we say that the star has contracted to a volume smaller than its "event horizon". Anything that has fallen into its event horizon can no longer be seen by external observers. Thus a black hole is formed.

The event horizon is the surface of the black hole. For a star whose mass ten times that of the sun, the radius of its event horizon is about 30 km. That is to say, when the radius of the collapsing star diminishes to 30 km, it forms a black hole.

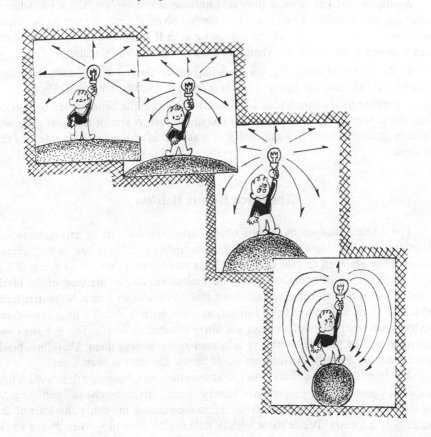

Fig. 9–2

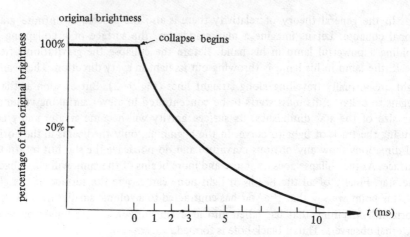

Fig. 9–3 The change of brightness of a collapsing star

Anything that has entered the event horizon is lost forever. Moreover, when a collapsing star shrinks into its event horizon, all physical processes will be powerless to arrest further collapse. It will be collapsing until it is reduced to a point at which many quantities go to infinity. Therefore the point is called a "singularity".

In the course of collapsing, the star will grow darker and darker, since the beams of light that can leave the star are becoming fewer and fewer in number. Fig. 9–3 shows the change of brightness of a collapsing star. From the figure it may be noticed that the process of its growing dark is incredibly fast. A star of ten solar mass will become almost invisible in a hundredth of a second after the commencement of the collapse.

The Black Hole is Hairless

The finite collapse of celestial objects gives rise to stars of various complex structures, yet the black holes resulting from infinite collapse are all very simple things. They are even simpler than any object we have ever seen. As we know all objects are made up of complex atoms and molecules. But in the case of the black holes, we need not, and cannot, talk about their molecular structure. No matter from what the black holes have been formed, as soon as they shrink within their event horizons we no longer need, nor can we, worry about their details. This is beause we can no longer be provided with any information concerning them. Therefore, black holes produced from different things would be just the same in most aspects.

Yet how simple is a black hole? To answer this one theorem suffices: a black hole possesses only three properties, namely mass, electric charge and angular momemtum. Once these three parameters are ascertained, the entire character of the black hole is known. Beside these a black hole has no other properties. In the whole universe, the entire character of no object other than the black hole can be presented by only three physical quantities. To the black hole we cannot ascribe any other property, just as we can say nothing about a patch of arid land. Therefore some people like to phrase the theorem in the following words: "A black hole has no hair."

According to this theorem, in the universe only a few types of black holes exist, the full exhibition of which can be seen in the following table:

name	type	property
Schwarzschild Black Hole	only has mass; angular momentum and electric charge are zero	the simplest spherically symmetric black hole
RN Black Hole	has mass, electric charge and zero angular momentum	spherically symmetric black hole with electric charge
Kerr Black Hole	has mass, angular momentum and zero electric charge	axially symmetric rotating black hole
KN Black Hole	mass, electric charge and angular momentum all differ from zero	axially symmetric rotating electric-charged most complex black hole

Here one point should be stressed: none of these black holes can have a magnetic field with its magnetic axis differing in direction from its rotation axis. The structure of a slanting magnetic field seen in the previously discussed neutron star (Fig.9—1) cannot possibly exist in the black hole.

Critical Mass

In the preceding passages we have repeatedly said that low mass objects will become neutron stars through finite collapse while the higher-mass objects will become black holes through infinite collapse. The limiting mass is called the critical mass.

According to the general theory of relativity, it can be calculated that the critical mass is about 3.2 times the solar mass.

Here let us sum up the relevant results:

1. A star whose mass is less than 3.2 times that of the sun will form neutron star, etc, which may have an inclined magnetic field.

2. A star whose mass is greater than 3.2 times that of the sun will form a black hole which cannot have an inclined magnetic field.

These are the principal theoretical conclusions concerning the outcome of collapse.

How do we test these predictions?

X-Ray Binaries

Let us first discuss the observation of black holes. A black hole itself cannot emit any light, but when some external object falls into its vicinity, due to the stress of the extremely high gravitational field, the object may copiously emit photons, their energies being in the X-ray or even γ ray spectral region.

Of course very few external objects fall into the black hole which is isolated in the sky. But among the celestial objects there are so many binaries each of which consists of two stars revolving round each other. During a certain stage of evolution this kind of systems will experience large-scale exchange of matter, i.e., the matter from one star will fall onto the other (Fig.9-4). In this manner, if the latter is a

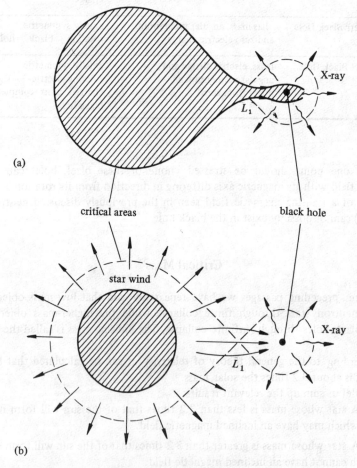

Fig.9-4 X-ray from close binary
(a) The matter of a star fills the critical areas
(b) The matter-exchange caused by the star wind

black hole, we may have the chance of detecting it, since this system should be an X-ray binary.

Since the 70's, artificial satellites or rockets have been used to make observations from beyond the atmosphere, and as a result many X-ray binaries have been discovered. According to the peculiarities of the intensity variation of the X-ray, it may belong to one of the two categories:

1. The intensity of the X-ray varies pulsatingly, its pulsating period being very stable (Fig.9-5).

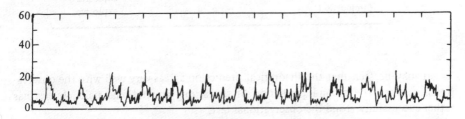

Fig 9-5 The rhythmic X-ray intensity variation of Hercules X-1

2. The X-ray intensity record shows explosive changes. The intensity change contour shows numerous irregular features and in no definite period can be identified (Fig.9-6).

From the knowledge of the radio-pulsar we know that a pulsating radiation is the emission of a neutron star which has an inclined magnetic field. The black hole cannot produce a pulsating intensity change with a stable period, since it cannot possibly possess an inclined magnetic field.

From the above discussion we see a way of testing our theory, which predicts that:

1. The mass of a pulsating X-ray source should be less than 3.2 times that of the sun.

2. The X-ray emision from a source of mass greater than 3.2 times that of the

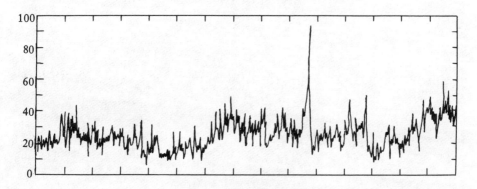

Fig. 9-6 The irregular X-ray intensity variation of Cygnus X-1

sun cannot possibly show pulsations

These two principles can be tested by observation. The following table shows recent observational results:

name of the X-ray binary	the mass of the X-ray source (in sun masses)	type
Centaurus X-3	0.7 ± 0.14	periodic
Hercules X-1	1.3 ± 0.21	periodic
Cygnus X-1	> 5	explosive
Circinus X-1	~ > 4	explosive

It may be seen that the theoretical prediction agrees very well with the observational results. The second prediction of the strong gravitational field physics has rather successfully stood the observational test.

CHAPTER X THE CONFIRMATION OF GRAVITATIONAL WAVES

Einstein's Prediction

The problem to be discussed in this chapter cannot be found in Newton's gravitational theory. The most essential difference between Einstein's field equation and classical gravitational theory is that only the former predicts gravitational waves.

What are gravitational waves?

Here let us explain by analogy. Fig.10–1 (a) shows a system that consists of two electrically charged objects. When oscillation takes place between them, electromagnetic waves will be emitted—this is one of the most fundamental conclusions of the electomagnetic theory. Fig.10–1 (b) shows a system that consists of two objects with masses. According to the general theory of relativity, when oscillation takes place between them, gravitational waves may be emitted.

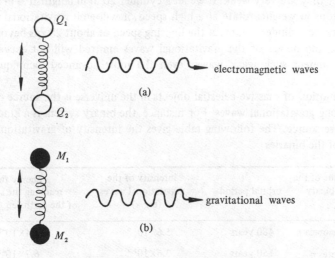

Fig. 10–1 (a) Two oscillating electric charges emit electromagnetic waves.
(b) Two oscillating particles emit gravitational waves

The propagation velocity of the gravitational waves is the same as that of light. Moreover, they carry a certain amount of energy with them. Therefore, they are a kind of real waves. Gravitational waves can be emitted and received.

All these features resemble very much those of electromagnetic waves. Therefore, though Newton's theory does not contain the conception of gravitational waves, this prediction of the general theory of relativity can be readily accepted, unlike the case of the black hole, the acceptance of which is rather difficult. Though the classical mechanics has already embodied the black-hole-type result, yet the black-hole conception of the general theory of relativity cannot be so easily accepted.

Though people can easily accept the prediction of gravitational waves, to confim them through observation is unusually difficult. Other predictions of Einstein in accordance with his general theory of relativity have all been verified through observation in a not very long period of time. Only the prediction of gravitational waves remained unproved for sixty years until towards the end of 1978, the first quantitative observational evidence was obtained.

The reason for the long lapse of time consists in the fact that the gravitational waves are really very weak.

Gravitational-Wave Source in the Universe

All accelerated objects emit gravitational waves. A little bouncing ball, the swinging arms of a man, the moon revolving round the earth ... they all emit gravitational waves. Yet they are very weak. If we let a cylinder 20 m in length, 1.6 m in diameter and 500 tons in weight, rotate at a high speed, it will emit gravitational waves. But, even when the cylinder rotates at the limiting speed of about 28 rps beyond which it will break, the power of the gravitational waves emitted will not exceed 2.2×10^{-19} watts. To detect such weak waves, even the most advanced techniques will not suffice.

The motion of massive celestial objects in the universe is the source of comparatively strong gravitational waves. For instance, the binary system is a kind of gravitational wave source. The following table gives the intensity of gravitational radiation of some of the binaries

name of the binary	orbital period	intensity of the gravitational waves (erg/s)	energy flux reaching the surface of the earth (erg/cm² ·s)
Cassiopeia η	480 years	5.6×10^{10}	1.4×10^{-29}
Bootes η	150 years	3.6×10^{12}	6.7×10^{-28}
Sirius	50 years	1.1×10^{15}	1.3×10^{-24}
Lyra β	13 years	4.9×10^{28}	3.8×10^{-15}
Leo UV	14 hours	1.8×10^{31}	3.5×10^{-12}

We see that though the gravitational radiation from binary is far more intence than that from the 500 ton cylinder, but compared with stellar electromagnetic out put it is extremely weak. For instance, the electromagnetic radiation of the sun reaches the power of 4×10^{33} erg/sec which is far greater than any in the above table. As for the energy flux that reaches the earth, it is still smaller.

Weber's Experiment

The first man that made an attempt to receive the gravitational radiation in the universe is the American scientist J. Weber. He devised and installed an antenna that can receive the signal of gravitational waves.

The way to receive gravitational waves differs from the way electromagnetic waves are received. It is very easy to receive electromagnetic waves. The human eye, the photographic plate and the radio set are all receivers of electromagnetic waves. The basic working principle is the same: the electromagnetic waves acting on the electrons set the latters in motion and the waves are detected through the effects of this motion.

Incontrast the peculiarity of the gravitational waves lies in that they turn and twist an object. Fig. 10-2 shows a round object, which, when gravitational waves are incident on its flat faces, will become an oscillating ellipsoid.

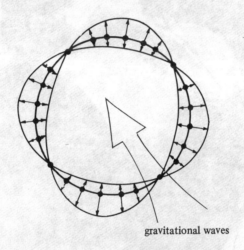

gravitational waves

Fig. 10-2 When the gravitational waves are radiating on its round surface, the round object will become an oscillating ellipsoid

Weber's gravitational wave antenna is an aluminum cylinder weighing 3.5 tons. On the surface of the cylinder a number of piezoelectric transducers crystal are installed which can respond to the slightest distortion of the cylinder. When gravitational waves reach the antenna, they are measured through the distortion of the cylinder.

The working principle of the antenna is very simple, but its manufacture involves a great deal of difficulty. Since many factors may cause the distortion of the cylinder, it is only by keeping away the external "noises" that may produce distortions that the gravitational waves can be observed.

In 1969 Weber declared that his antenna had twice received gravitational wave's signals, in his 81 days observations from Dec 30, 1968 to Mar 21, 1969.

The publication of Weber's experimental results attracted much attention of the physicist circles. In many countries gravitational wave experimental teams were set up to repeat his experiment. But Weber's results have also aroused a great deal of doubt.

Fig. 10-3 Weber and his gravitational wave detector

Firstly, if what Weber received were the gravitational wave's signals that came, as Weber declared, from the centre of the Milky Way, then at the centre of the Galaxy some very violent events must have taken place, yet when astronomical observational material was cheched, no extraordinary records whatever were found.

Secondly, if the energy density of the gravitational waves when reaching the earth was as great as Weber stated, i.e. 10^{10} erg/cm$^2 \cdot$s, then a mass of 10^4 suns should be consumed every year at the centre of the Galaxy, to generate such strong gravitational waves. If that were the case, the entire life span of our Galaxy could not have exceeded 10^7 years. Yet astronomical observation proves that our Galaxy has already had some 10^{10} year's history. This is another contradiction.

Moreover — and this is more important than the previous two arguments, —Weber's results have not been repeated by the experimental teams of other countries. Therefore Weber's results have not won general acknowledgement.

The general opinion now is that the sensitivity of the gravitational wave antenna in present-day laboratories is still too low to be able to detect the gravitational signals in the universe. Therefore to improve the antenna sensitivity through various methods is what the experimental teams are aiming at.

Gravitational Radiation Damping of a Binaries

Meanwhile the astrophysicists blazed a new trail to test the theory of gravitational waves.

As we have said before, the binary is a typical source of gravitational radiation. The gravitational radiation can slowly but surely carry away the energy of the system. As a result the period of the binary will become shorter, which phenomenon is called the gravitational radiation damping.

If only we can prove the existence of this phenomenon, we have found support to the theory of the gravitational radiation, even though we have not directly observed the gravitational waves. And this is the method the astrophysicists have adopted.

But this method can hardly bring forth the expected results either, since there are many additional factors that influence the period of the binary. For instance the mass exchange between the two component stars (Fig.9—4) can bring about a change in period of the binary. Besides, the tidal effect between the two stars can also cause a change. According to geophysical and palaeontological analyses, several hundred million years ago the period of revolution of the moon round the earth differed from what it is today, and this change has been caused by the tidal effect between the earth and the moon. Moreover, the stellar wind (the flux of particles flung out by the star) may also reduce the mass of the binary system, thus causing a change of the period.

All in all, the factors that affect the period of the binary may be grouped into two calsses. The first consists in gravitational radiation, damping being a relativistic effect. The other class is made up of nonrelativistic factors such as ocean tidal. The binary system that suits the purpose of testing the theory of gravitational radiation damping should satisfy the following requirement:

relativistic factor $\gg$ nonrelativistic factors.

According to the general theory of relativity, the gravitational radiation damping is inversely proportional to the fifth power of the distance a between the two stars in the binary (i.e. a^5). Therefore, for the sake of measuring the relativistic effect, we should choose those binaries with shorter distance between the pair. On the other hand, the tidal effect is directly proportional to

$$(\frac{R}{a})^3$$

where R is the radius of the star. Therefore, in order to reduce the nonrelativistic factor, a greater distance between the stars is desirable.

These two requirements are contradictory to each other. Therefore a binary system that consists of such stars as the sun can never satisfy both conditions.

From the above it can be seen that only when the radius R of the star is sufficiently small, can the nonrelativistic factor be greatly weakened so that the relativistic factor will dominate over its opponent factor.

Therefore only binary systems consisting of two compact stars (R very small) can serve as a good celestial laboratory to test the theory of gravitational waves.

Not until 1974 was any binary found that consists of two compact stars.

PSR1913+16—An Ideal Relativistic Celestial Laboratory

Towards the end of 1974 the American radio astronomers R. Hulse and J. Taylor discovered a radio pulsar—PSR1913+16[1] as it is called. The pulsar differs from all others that had been found by then in that they are single stars while PSR1913+16 alone is definitely a member of a binary system.

The pulsating period of this pulsar is very short: it is only 59 milliseconds. Among the periods of all discovered radio pulsars, it only exceeds that of the Crab Nebula pulsar (see Chapter IX). Moreover, the period of the binary system is also very short (less than 8 hours), yet the orbital eccentricity is very large. The combination of all these characteristics in one case is a very rare thing. The pulsar has attracted wide attention.

In this binary system we can only observe PSR 1913+16. What about the other star of the system? For it we can only make some conjectures based on the evolution of the system.

The evolution of a binary system roughly undergoes five stages as shown in Fig 10–3. In the first stage they are still two common stars. In the second stage the one of greater mass begins to expand, so that its matter flows incessantly to the other one of smaller mass. In the third stage the star of greater mass experiences a supernova explosion and becomes a compact star consequently. In the fourth stage the star of smaller mass begins to expand and its matter continually flows to the compact star. By this time the binary system should be an X-ray binary. In the fifth stage, the star originally of smaller mass undergoes a supernova explosion and the system then contain two compact stars.

During the supernova explosion, a great deal of energy is released, which usually brings about the disintegration of the binary system. For this reason a binary system rarely contains a compact star. The formation of a binary star system entails two supernova explosions. It cannot but be a very rare case that these two explosions should have not disintegrated the system. That is why the double compact star system is a rarity in nature.

---

(1) PSR, abbreviation of pulsar;1913, its right ascension; 16, its declination.

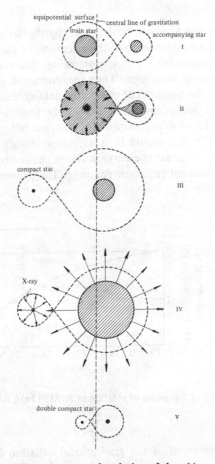

Fig. 10-4. The main stages of evolution of close binary

Judging from the property of PSR1913+16, the invisible companion star should also be a compact star. Since PSR 1913+16 does not emit X-ray, it cannot be in the fourth stage. Moreover, the distance between the two stars of the system is very small and thus the system cannot be in the third stage (i.e. the other star is not a compact one). The first and the second stages do not contain any compact star and can therefore be excluded from consideration. Hence there remains only one possibility, namely that PSR1913+16 is in a system of double compact star.

In the middle of the 70's PSR1913+16 was the only known system of double compact star. Right now it is still the only sky laboratory fit for the purpose of testing the gravitational wave theory.

Gravitational Radiation Damping Tested

According to the general theory of relativity, the shortening of the revolution period of a system of the two compact stars is mainly due to gravitational radiation damping. Therefore the very confirmation of the shortening of the period of PSR 1913+16 is an unequivocal evidence of the soundness of the theory.

Taylor and others carried out a monitoring observation of PSR1913+16 for more than four years in succession. Their measurements amounted to more than a thousand times, and the precision of observated data exceeded one part per thousand million. These data really confirmed the steady shortening of the period of the binary system. Fig.10—4 shows its phase of rotation relative to time. If there is no period shortening, the line should be a horizontal straight line. The dots represent observational results, whereas the curve is the calculated results in accordance with the theory of gravitational radiation damping. The agreement between theory and observation is apparent.

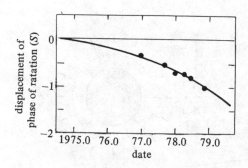

Fig. 10—5 The phase of rotation of PSR1913+16 relative to time

The quantitative proof of the gravitational radiation damping is of great significance. Once again observation convincingly proves the correctness of the general theory of relativity. The success greatly furthers the advance of gravitational physics.

This success, like the discovery of the neutron star, is the product of the ingenious combination of various works. Viewed from the theoretical angle, it involves the general theory of relativity, the theory of evolution of the binary and the computation of various periodic changes. As for the practical measurement, the largest radio-telescope of today (300 m aperture) has been employed in addition to precise time keeping and X-ray and optical measurements.

If the 19th century discovery of Neptune is the most brilliant observational proof of Newton's gravitational theory, then this 20th century confirmation of gravitational radiation damping might be regarded as the most brilliant observational proof of relativistic gravitational theory.

CHAPTER XI FROM NEWTON'S UNIVERSE TO THE EXPANDING UNIVERSE

From the Finite Bounded to the Infinite Unbounded

In this chapter we shall turn to the last and also the grandest physical problem, i.e. cosmology.

In the development from classical mechanics to the theory of relativity, what changes have taken place in the human cognition of the physical structure of the universe?

Before Galileo and Newton, the traditional conception of the structure of the universe can be illustrated in Fig. 11–1. That is a finite bounded world. The outermost layer of the universe is made up of the sky of fixed stars, beyond which no space exists. It is the boundary of the universe in the Copernican theory that maintains the sun being the centre, the finite bounded structure is preserved.

After Newton, the view of an infinite unbounded univese has been universally adopted. That is to say the volume of the universe is infinite, and has no spatial boundary. The space of the universe is an infinite Euclidean space of three dimensions, i.e. in the directions up and down, left and right, forward and backward, space can extend to an infinite distance.

In this Newtonian infinite box, celestial objects are scattered everywhere, and the number of the celestial objects are also infinite. No matter in which direction we proceed, we shall see no end of them.

All in all, the infinity of the space of the universe, just like the universality of the Newtonian theory, has been accepted as a matter of fact in classical physics.

Of course, the natural philosophy of infinite world had played a decisive role in breaking the spiritual fetters of the religious cosmology of the Middle Ages. The scientific revolution represented by Copernicus, Galileo and Newton has once for all overthrown the notion of the earth being the centre of the universe. The very memory of this revolution evokes endless admiration.

But the awe-inspiring fame that accompanies every great success often so affects the successors that they may refuse to consider, or dislike to consider what among the consequences is really the proved truth and what remains a conjecture or a hypothesis. In fact, that the space of the universe is a Eucldean space of three dimensions and that Newton's theories can be applied to cosmology are two ideas pertaining to the second category. Though people are in the habit of taking these ideas for granted, they are by no means proved truth.

The Difficulty of Newtonian Infinite Universe

In the development of relativistic cosmology, the first step Einstein took was to point out the contradiction and inconsistency that abide in the Newtonian concep-

Fig. 11–1 The Ptolemiac structure of the universe

tion of infinite universe.

The way the Newtonian mechanics discusses the motion of a finite mechanical system is based on the supposition that a frame of reference can be adopted in which the gravitational potential φ becomes a constant in infinitely remote place. This condition is rather crucial in solving problems of motion of celestial objects in a limited region. Yet by accepting the Newtonian picture of an infinite universe with a homogeneous distribution of matter, we necessarily come to the conclusion that the gravitational potential φ, according to Newtonian mechanics, cannot be a constant, which assertion is contradictory to the first supposition. On the other hand, in order that φ may be a constant in infinity, we may expediently abandon the supposition that matter is distributed homogeneously through out the entire infinite space, believing that it is mainly concentrated in a finite region surrounding us. Thus reasoning, we see that though φ becomes a constant at infinity the material world cannot be but finite.

Therefore Newtonian mechanics in principle cannot be employed to describe (not to say accurately describe) the kinetics of the physical system of infinite universe. Either the Newtonian theory, or the concept of infinite space, or both, should be corrected. This is a "simple" yet most essential issue that Einstein offered to cosmology.

An "Idiotic" Problem

Yet is the problem that Einstein has brought forward of any meaning?

One of the arguments is that, the universe being so large and so complex, to discuss it as a kinetic physical system may bear no significant fruit.

Yet a "solution" of this kind cannot satisfy Einstein in the least. "If I am asked to abandon so much in this task of principle, I shall feel very sad. Unless all these efforts for a satisfactory comprehension have been proved to be vain and useless, I'll not make such a decision." Einstein never made such a decision. In fact he always held a resolute attitude towards his work. He firmly believed that in the universe there must exist some highest universality. He always cherished "an admiration, a devout feeling, not towards mankind, but towards the mysterious harmony of Nature wherein we are born."

When Einstein undertook to solve this problem, he wrote a letter to W. de Sitter, saying: "Is the world infinitely extensive or finitely closed? In answer to this question Heine in one of his poems says: 'Only an idiot expects a solution'." Really, many problems that have captivated the fancy of the physicists and astronomers are next to absurdity even in the eyes of the most imaginative poets. It seems only idiots are willing to waste their energy on such problems. Yet the fact is not like that. As regards natural science, we can say there is no problem concerning nature but merits our study. Our present world abounds, not with "idiotic" problems, but with idiotic answers. When people say a question is "meaningless", they are probably giving an answer of this kind.

There is still another reason why Einstein felt it necessary to study the problem: at that time it was only cosmology that contained the problem of strong gravitational field. If celestial objects, as Newton's classical conception asserts, are more or less homogeneously scattered throughout the whole space, their average density being ρ, then the total mass M in a sphere of diameter R is roughly ρR^3.

For this sphere

$$-\frac{GM}{c^2 R} \approx \frac{G}{c^2} R^2$$

from which it is apparent that when R is very big, the value will certainly approach 1. Therefore when treating the cosmological problem, in principle we cannot employ Newton's theories. In this problem, the general theory of relativity is not a small correction of the Newtonian theory (as that in the post-Newtonian problem): it brings about a fundamental change of Newton's theory.

The Finite Unbounded Universe

The fundamental difference lies in that the general theory of relativity refutes the *a prior* supposition that the space of the universe must be a three-dimensioned boundless Euclidean space, since the structure of the space of the universe is not independent of the motion of matter in the universe.

The first model of the universe given by Einstein is neither Aristotle's finite bounded system nor Newton's infinite boundless system, but a finite system without boundary. When we say finite we mean the finite volume of space. When we say without boundary, we mean that this three-dimension space is not a part of a greater three-dimension space: this space has already included the entire space.

As is shown in the history of cosmology, the conception of a finite thing without boundary made its first appearance not in Einstein's model of the universe. We have already mentioned in the first chapter that Aristotle insists that the earth is not a boundless plain but a sphere. In fact, this is to replace the structure of an infinite yet unbounded plain by that of a finite unbounded sphere. The spherical surface is a two-dimension finite system without boundary. Traversing along the surface of the sphere one can never reach its end.

If we only carry Aristotle's conception of two-dimensional finite boundlessness to three dimensionsal, we shall obtain Einstein's three-dimensional finite unbounded system. In various aspects these two conceptions are analogous. For instance, the spherical surface is a two-dimensional curved surface, while a finite unbounded three-dimensional space is a curved space.

When we say "curved", we mean to deviate intrinsically from the Euclidean geometry. For instance, as to a spherical surface (two-dimensional), we shall measure it in the following way. As Fig. 11–2 shows, starting from point A we move along the big circle to B. The length between A and B is called R. Then we draw a circle (one-dimensional) with A as the centre and R as the radius. The length of this circle is l. In Euclidean geometry

$$\frac{l}{R}+2\pi$$

but in the case of a spherical surface

$$\frac{l}{R}<2\pi$$

Hence the spherical surface is a curved surface, instead of a plane discussed in plane geometry.

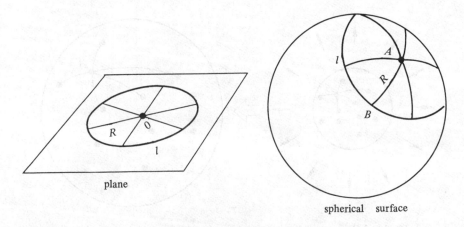

Fig. 11–2 The spherical surface is a curved surface. On the spherical surface geometry differs from that on the plane (Euclidean geometry)

By analogy, in carrying the two-dimensional spherical surface to a finite three-dimensional curved space withoutboundary., we shall make measurement in the following manner: starting from point A we move to B, the length from A to B is called R. Then, just like in the previous example, we shall carry the one-dimensional circle to two-dimensional spherical surface with A as the centre and R as the radius. The area of the spherical surface is S. In Euclidean geometry

$$\frac{S}{R^2}=4\pi$$

while in Einstein's finite unbounded model, we have

$$\frac{S}{R^2}<4\pi$$

That is to say in Einstein's model, the inconsistency of the Newtonian system no longer exists. Of course, the absence of inconsistency is only a prerequisite, but not a sufficient condition, of a correct theory. The important test awaiting a theory is the contrast between the theory and the observation.

The Expanding Universe

Following Einstein's first model, other models were proposed by other cosmologists. Among them A. Friedmann and G. Lemaitre separately advanced the model of an expanding universe. When we talk about an expanding universe, we mean the characteristic length scale of the universe is continually increasing with time. If we still use a two-dimensional finite unbounded spherical surface as the analogy to a three-dimensional finite unbounded system, then an expanding two-dimensional sphere will be moving as Fig. 11–3 shows.

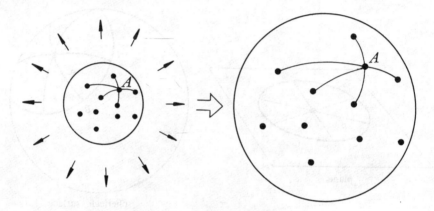

Fig. 11–3 On an expanding spherical surface the distance between any two points is growing greater and greater

The small dots in the picture represent material objects on the spherical surface. It is obvious that when the spherical surface is expanding, the small dots will become more and more sparse, the distance between any two of them grow larger and larger. Now let us imagine an observer standing on one of these dots. He will discover that all the other dots are moving away from him. Moreover, those small dots nearer to him move at lower speeds, while other dots farther away move at higher speeds. The farther the separation between these two dots, the higher is the speed at which they are leaving each other.

In the year 1929 the American astronomer E. Hubble discovered that the spectra of all the external galaxies show red shifts. Red shift is the increase of the wavelengths (or the decrease of frequencies) of spectral lines. If the wave length of a spectral line produced originally by a certain atom is λ_0, then the wavelength λ of the spectral line originated by the external galaxy is generally greater than λ_0. As a rule we use $z = \frac{\lambda - \lambda_0}{\lambda_0}$ to represent the magnitude of the red shift, z being termed simply as the "red shift".

From the characteristics of this kind of red shift, we may believe that it is due to the Doppler effect. What is called Doppler effect is that when the light source moves relative to the observer, the wavelength of the light received by the observer differs from that when the source is at rest. As Fig 11–4 shows, a light source A moves towards the observer while another light source B moves away from him. If the wavelength of both source A and source B are the same, i.e. λ_0, the wavelengths of these two beams as the observer receives them will be different. In his eyes λ_A of source A is smaller than λ_0, while λ_B of source B is greater than λ_0. We usually say the wavelength of source A has blue-shifted while the wavelength of source B has red-shifted. The higher the relative velocity, the greater the red or the blue shift.

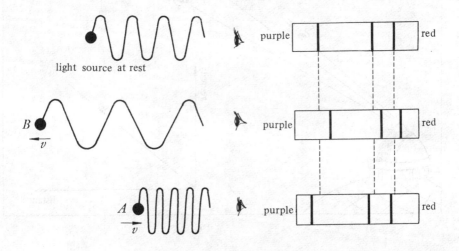

Fig. 11–4 Doppler effect

To explain the extragalactic red shift by the mechanism of Doppler effect signifies the receding of the galaxies from us. Hubble also discovered that the amount of the extragalactic red shift is related to how far the galaxy is away from us. The nearer the galaxy, the smaller the amount of the red shift, and vice versa (see Fig. 11–5). This property is generally referred to as the Hubble relation. Translated into ordinary language, Hobble relation tells us that the farther the external galaxy is away from us, the faster is is moving away.

The galaxies studied by Hubble all exhibit a small amount of red shift: in all cases $z<0.003$. In the following decades a great number of galaxies have been found that show much greater red shifts. Among the known galaxies, the greatest red shift reaches to $z\approx1$. In this enormous range Hubble relation still holds. Uo to the present all observational results agree with the prediction of the expanding model.

The conception of an expanding universe radically alters the traditional conception of cosmology which maintains that viewed on a "large scale" the celestial objects

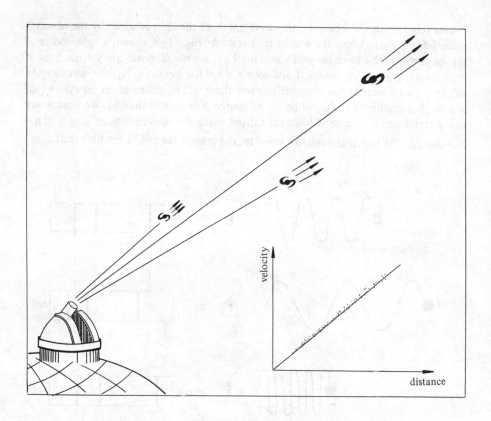

Fig. 11-5 The farther away the galaxy the faster it recedes from us

should be at rest, though the sun and the celestial objects in a "small" region as our Milky Way are in motion. That is to say viewed from the angle of a still "larger" scale, the mean velocity of the celestial system should be zero. The formation of this conception may be explained by the fact that of the celestial panorama viewed with our naked eye, except that the stars rise in the east and fall in the west, other changes can seldom be witnessed. Even Einstein himself was not free from the bondage of this traditional conception, though the solution of his gravitational field equation can only be a universe in motion. Since he felt that motion on a large scale was unacceptable, he tried hard to produce a stationary model even at the expense of correcting his gravitational field equation. Since the discovery of the galactic red shift he felt sorry for what he had been attempting: the expansion of the universe is a natural corollary of his general theory of relativity, yet he had once tried to reject it. Einstein said afterwords that the attempt was " the biggest blunder" in his life.

The Big-Bang Cosmology

If the universe is expanding, then previously the universe should be smaller and matter denser. It follows that in its early epoch the universe might have been very compact. The density of matter was incredibly high, entirely different from what we see today in the starry world.

Following this clue in our study of the history of development of the universe, we attain what is called the big bang cosmology. At present the hot big bang cosmology is most popular.

The main view point of this school is that our universe has history of evolution from dense to rare and, from hot to cold state. More concretely, just over 10 thousand million years ago a big bang took place. At that time the density of matter of the universe was greater than even that of the nucleus; the temperature was also very high, reaching more than a million million degrees. During the early stage of the explosion, matter in the universe took the form of various kinds of particles, such as neutrons, protons, electrons, photons, neutrinos, muons, π mesons and hyperons. These particles incessantly collided with each other and transformed from one into the other. The entire universe was basically in a state of thermodynamical equilibrium. For instance an electron and a positron collided and were annihilated, to give birth to a pair of γ photons. Similarly the interaction of the γ photon gives birth to pairs of electons and positrons. In a second several million million million times of transformation might have taken place, each action being balanced by a counteraction. That is the very early epoch of the universe.

This very early epoch was very short, perhaps less than one minute. Because of the continuous expansion of the entire system, its temperature went down abruptly. Then the evolution of the universe passed into another stage during which neutrons began to lose the conditions required for their independent existence. They would either disintegrate or be united with protons to form such elements as deuterium and helium. The chemical elements in the universe began to be formed in this period which lasted about 30 minutes, temperature being about a hundred million degrees.

During the high temperature epoch of several hundred thousand years, the radiation in the universe was very intense. Thermal radiation and other particles were in a state of equilibrium. After this epoch, the density of matter decreased. Only after the temperature had lowered to several thousand degrees did the effect of the thermal radiation on other particles began to decrease in a large measure, and thermal radiation began to propagate freely without being affected by matter. With the expansion of the universe the temperature of this free thermal radiation was gradually decreasing, though it still maintained its original peculiarities[1].

---

(1) At a certain temperature, if thermal equilibrium is attained, the intensity of the thermal radiation is distributed in accordance with its different frequencies. The radiation characterized by this kind of frequency distribution is called thermal radiation.

From the time when the interaction between thermal radiation and other matter had become negligible until the present, more than 10 thousand million years have elapsed. In the history of evolution of the universe, this epoch was the longest. At the initial stage of this epoch, matter in the universe was mainly in the form of gas, from which gradually originated nebulae, By and by the nebulae shrank into galaxy star clusters, stars, planets...until they took the form of the starry world that we see today.

The following picture shows the main stages of evolution of the universe as described by the hot big bang cosmology.

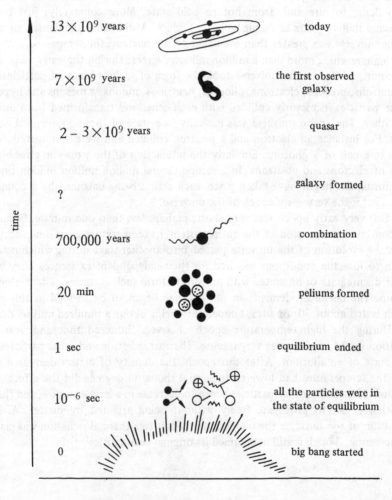

Fig. 11-6 The main stages of evolution of the universe described by the hot big bang cosmology

Yet what facts give support to the hot big bang cosmology?

The Age of Objects

The first fact that supports the hot big bang cosmology is the age of the celestial objects. Since big bang cosmology maintains that just over 10 thousand million years ago no star existed in the universe, the age of all stars, should be less than the age of the universe, i.e. some 10 thousand million years. Observation supports this argument.

One of the methods by which we measure the age of a celestial object is the use of radioactive isotopes. For example uranium exists in two isotopes, ^{235}U and ^{238}U. Both are radioactive, but their half lifes are different. The half life of the former is seven hundred million years, whereas that of the latter is four thousand five hundred million years. Since ^{235}U degenerates much faster, with the passing of time, the amount of ^{235}U will become lesser and lesser in comparison with ^{238}U. By the ratio of the amount of ^{235}U and that of ^{238}U, the age of the celestial objects can be estimated. By isotope chronology, we find that the age of the solar system is about 4.5 billion years, and the element of uranium in the solar system was produced between 5 to 11 billion years ago.

Another age-ascertaining method is through the investigation of the globular star clusters. A globular cluster is a system made up of nearly a million stars. We can measure the luminosity and the surface temperature of each and every star in the system. With these data we can draw a diagram, the horizontal axis of it being the surface temperature, the perpendicular axis the luminosity. Putting down the points

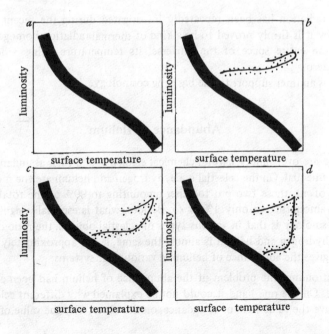

Fig. 11–7 The Hertzsprung-Russell diagrams of the globular clusters

that represent the stars of the cluster on the diagram, we shall discover that with respect to different globular clusters, there are different shapes of distribution (see Fig.11–7) which, according to the theory of star evolution, in fact signify different ages. The diagrams in Fig.11–7 are literally arranged in order of age. By the use of these diagrams we can ascertain the age of each of the globular clusters. The oldest of them are all aged between 9 to 15 billion years.

None of these results contradicts the requirement of the big bang cosmology.

Microwave Background Radiation

The big bang theory also predicts that some thermal radiation of the early epoch of the universe should still be around. It marks the temperature of the universe.

In 1965, A. Penzias and R. Wilson of the Bell Telephone Laboratories were engaged in installing a ground station for communication satellite when they found that some "noise" of unknown cause was interfering the reception and that they could do nothing to eliminate it. Their antenna was operating at 7.35 cm wave length. Later on the news reached some astrophysicists in Princeton University. The Princeton people decided that this was the cosmic radiation predicted by the hot big bang theory. Since the radiation permeates the whole space, it constitutes the indelible "noise".

This radiation has been repeatedly investigated during the recent ten years or more. Now it is firmly proved to be a kind of thermal radiation homogeneously permeating the entire space of the universe, its temperature being 3 degrees above absolute zero.

This is another support to the big bang cosmology.

Abundance of Helium

There are over ninety natural chemical elements, and their abundances in nature are quite unequal. On the celestial scale, hydrogen and helium are the most abundant elements of all, these two put together amounting to 99% of the total mass, other elements amounting to only 1% or so. Besides, what is especially significant in the eye of cosmology is that in various types of celestial objects, the ratio of abundance between hydrogen and helium is almost the same, ie 3:1 approximately. The following table gives the abundance of helium in various star systems:

In astronomy the problem of the abundance of helium had been a puzzle for a long time. On the one hand it could not be explained why different celestial objects should have the same helium abundance; on the other, why the value of it is roughly 30%.

The big bang cosmology can quantitatively explain the helium abundance. It was

Star System	helium abundance
The Milky Way	0.29
The Small Magellanic Cloud	0.25
The Large Magellanic Cloud	0.29
M 33	0.34
NGC 6822	0.27
NGC 4449	0.28
NGC 5461	0.28
NGC 5471	0.28
NGC 7679	0.29

during the first few tens of minutes in the early epoch of the universe, the efficiency of the helium production was very high. According to the measurement of the expanding velocity of the universe and the temperature of the thermal radiation, we can calculate the abundance of helium created in the early epoch of the universe. The calculated datum is exactly 30%. That is to say, the 30% helium that we find today in various celestial objects is probably the trace left behind by the big bang that took place more than 10 billion years ago.

The big bang cosmology is a developing school. Besides the aforementioned successes, there are a series of problems yet to be solved. However, after all these meticulous experiments and serious meditations that marked the progress from classic cosmology to modern cosmology, today we even possess a capacity to decide many things that happened more than 10 billion years ago. This may very well be regarded as a great triumph of the power of human knowlege.

CHAPTER XII AFTER EINSTEIN

Seeking After Unification

When we recall the scientific progress from Aristotle to Newton and then to Einstein, it seems that we can identify a persistent drive of the physical science: it ever endeavours to find unified law that governs diverse processes, and a unified origin out of which all types of matter were born.

A long time ago Aristotle proposed that our multifarious world originated from a single substance which he called *ylem*. This, however, is only a philosophical conjecture. The first unity in the scientific sense was the gravitational law discovered by Newton, a law that governs the motion of not only celestial objects but also of falling objects near the surface of the earth. This we have already mentioned in the first chapter.

The second great leap forward was completed in the nineteenth century by J. Maxwell who established the electromagnetic theory, thus attaining a unity of electric, magnetic and optical phenomena.

After founding the special and general theories of relativity, Einstein expended all the energy of the remaining half of his life seeking the unity of gravitation and electromagnetism. Once he said:

"... It cannot be asserted yet that those parts of the general theory of relativity that have been regarded as confirmed truth have provided for physics a complete and satisfactory basis. First of all the general field that appears in it consists of two logically irrelevant parts, i.e. the gravitational part and the electromagnetic part. Secondly, like the previous field theories, this theory up to the present has not yet offered an explanation concerning the atomistic structure of matter."

This is Einstein's view towards the unified field. Yet Einstein died without having found a logically unified general field. For a certain period of time some people failed to fully comprehend Einstein's thought on a unified field, assessing it in a completely negative spirit. Of course it is never difficult to shake one's head at anything not yet accomplished, presuming it to be a failure after all. Yet the question is how to assess the thing properly. In recent years, inspired by development in particle physics, the search for a unified field has once again become a principal orientation of theoretical work. For this event can be traced back to the sixties of this century.

The Great Unity and the Very Early Universe

Towards the end of the sixties, people realized that in the universe all the physical phenomena can be grouped into two classes: the class of "matter" and the class of "interaction". The former includes things such as quark, electron and neutrino, the latter includes gravitational and electromagnetic forces. In the present universe, the fundamental interaction can be classified into four types (in order of intensity): the strong interaction in which hadrons partake the electromagnetic interaction in which the electrically charged particles participate; the weak interaction in which both hadrons and leptons participate; the gravitational interaction, which is the weakest of all, in which all kinds of particles participate. Yet up to that time each of these interactions had developed its own theory independently: interrelations between these theories were wanting. Among them the theory of strong interaction and that of weak interaction are most dissatisfactory, because computation based upon these theories leads to various infinities. Many attempts have been made to eliminate these difficuties, but all to no avail.

In 1960's the situation somewhat improved. S. Weinberg, A. Salam and S. Glashow once again set eyes on unity. They independently proposed a theory which unifies electromagnetic and weak interactions and which came to be called the weak-electro unified theory. This unified theory has not only solved the problem of infinity in the field of weak interaction but has won support from many experimental results.

This success inspired people to seek theory of even greater unity, unifying the strong interaction with electromagnetic and weak interactions—the grand unified theory as it is commonly called. At present there are many schemes for this grand unification, yet we cannot say which of them has been decidedly proved to be valid. The difficulties that attend the proof of the theory lie in the relevant experiment. The basic conception of the grand unified theory is that with the increase of energy the coupling strength of the strong interaction weakens, that of the electromagnetic interaction remains constant, and that of the weak interaction varies. These coupling strengths will come equal only when the energy is about 10^{24} eV (approximately 10^{12} erg).

10^{24} is so fabulously high that we can never hope to carry out this kind of high-energy experiment through accelerator. The presentday accelerator has raised the centre-of-mass energy to about 10^{10} eV. Our next generation may expect an energy as high as 10^{12} eV. This energy means a great deal to Weinberg-Salam's theory of weak-electro unity, but is by far too small for the theory of great unity.

Yet where can such an enormous energy be found?

Perhaps only in the earliest epoch of the big bang of this universe wherein we live did there ever exist an energy of the order of 10^{24} eV that marked the particle process then going on. So the universe in its earliest epoch, or the universe whose age was younger than a wink of 10^{-6} second, might serve as a "laboratory" in which the high-energy behaviours required by the theory of great unity and the theory of weak-electro unity could have been testified. This is the immediate reason why in recent years the particle cosmology has rapidly developed.

One of the most interesting developments of particle cosmology is the explanation of the asymmetry between particles and antiparticles. Since a careful study of this problem will stray too far from the theme of the book, here only   most general relation of its import will be made. In 1928 Dirac established the relativistic quantum theory of electrons which was verified by the discovery of positrons in 1932. Since then people believe that both particles and their antiparticles exist in the universe, their respective properties being symmetrically opposite, their abundance in the universe being also symmetrical. Yet the astronomical observation has provided us with an opposite result. In the present-day universe, the number of particles by far exceeds that of antiparticles—the cosmic asymmetry of particles and antiparticles as it is called.

As to the origin of this asymmetry, the grand unified theory can give a fairly natural explanation. The main reason is that the grand unified theory predicts a process which breaks the symmetry between particles and antiparticles, and which in turn might bring about the unequal numbers of particles and antiparticles. This kind of process, however, plays a very insignificant role in the present-day universe and can only be witnessed in a few experiments. Yet this minute symmetry-breaking process, in the earliest epoch of the universe, when the age of the universe was less than 10^{-36} second, might have influenced the whole universe, making the number of protons greater than that of the antiprotons. Yet the quantity of this asymmetry at that time was very little, amounting to only a billionth part or so. It is only when the universe had cooled down, the inflence of it became more and more conspicuous. The earth and the entire starry world are made up of this very little amount of "surplus" particles left over by the primordial fire ball of the universe.

Though this explanation has not been confirmed by sufficient evidence, the very way of thinking that characterizes the explanation is nevertheless very tempting. This is because that ever since Newton the study of nature has split up into two branches, one is concerned with the structure of matter of ever diminishing size, the other with the phenomena of the universe whose scope is ever broadening. But now the exploration of the earliest history of evolution of the universe and the probing into the innermost depths of the structure of matter have begun to join together.

Following this line people naturally expect to go still further, seeking a more unified theory than the great unity theory, probing into a universe even earlier than the earliest epoch.

Gravitation and Quantum Theory

What awaits to be unified after the grand unification cannot but be the gravitational interaction. Gravitation is the earliest-known interaction, yet the problem of its unification is the hardest nut to crack. One of the reasons is that gravitational theory and quantum theory have not yet been reconciled to each other. Einstein held a negative attitude towards the basis of the standard interpretation of quantum theory. He believed that through the unification of the gravitational interaction and the other interactions a correct theory can be found to replace the present quantum theory. Yet this line has not proved to be promising. On the contrary, all the unifications that have proved successful have been accomplished within the framework of quantum-field theory. Therefore it is now generally believed that in so far as the relation between gravitational theory and quantum theory is concerned, Einstein's view should be turned upside down, i.e., only after finding a theory of gravitation compatible with quantum theory can unification be finally attained. Therefore, though within the limited range of energy in our present-day laboratories the quantum effect of gravitation has not yet been found, the search for the quantum-gravitation theory has ever maintained its pith and moment.

Yet the work has encountered a series of difficulty. People are beginning to be aware that the difficulty is probably not a technical one; it might involve some of the most fundamental conceptions. For instance, in the quantum theory of gravitation there is an energy scale of the order of 10^{28} eV. When energy exceeds this value, space-time itself can no longer serve as a continuous background against which the description of a progressive motion is made but will betray an obvious quantized rise and fall. Therefore beyond this scale it is impossible to talk about the deeper layers of the structure of matter. In other words the quantum theory of gravitation will demand a microscopic limit to the structure of matter.

The second fundamental problem of the quantum theory of gravitation is related with the law of cause and effect. As we know, in classical physics we can accurately predict both the position and the momentum of a given particle, whereas in quantum theory, in accordance with the uncertainty principle, we can only accurately predict either the position or the momentum of it, ie the predicting power has thus been reduced by half. This indicates a weakening of the law of cause and effect. In the quantum theory of gravitation the law of cause and effect will be further weakened. For instance, in the process of quantum emission by a black hole, we can only predict the probability of a certain mode of a certain type of particle. Yet the emission of the black hole is not the effect in the quantized theory of gravitation. The phenomenon can only be regarded as semi-classical and semi-quantum, about which a little more will be said by and by.

Black Hole Emission

In the year 1970 a theorem in classical black hole theory has been proved: the surface area of an evolving black hole can only increase and cannot decrease. When external matter or radiation falls into the black hole, the surface area of the event horizon of the black hole will increase accordingly. When two black holes collide to merge into one, the ultimate surface area of the event horizon will be larger than the sum of the former two areas. No matter can escape from within the black hole, and one black hole cannot split into two. Therefore the surface area of the event horizon of the black hole can only increase. This is called the theorem of undiminishable area.

That the surface area of the black hole can never decrease reminds people of the entropy in the thermodynamics which is also undiminishable in its progression. The entropy in an isolated system always increases with time (or remains constant), but never decreases.

Later on it was further discovered that the black hole area and the thermodynamical entropy are analogous not only in form but also in essence. The area is the real entropy of the black hole. As to the temperature of the black hole, we may so define it: this temperature should have the same significance as that in the thermodynamics. For instance, thermal equilibrium should be maintained between objects of the same temperature.

As we know, thermal equilibrium is a kind of stationary equilibrium. When objects A and B of the same temperature have reached the state of equilibrium, between them there may still be energy exchange. But in any time interval, the energy that flows from A to B is equal to the energy that flows from B to A; as a result the net effect is that the temperature of both A and B remains unchanged. In this manner, if thermal equilibrium can be attained between a black hole of a certain temperature and another object of the same temperature, then in a unit time the heat that has flowed from the object to the black hole is equal to the heat that has flowed the other way round. Yet the classic theory of black hole insists that not any matter whatever can escape from within the black hole. This is an inconsistency of black hole thermodynamics.

In 1974 S. Hawking solved this contradiction. The most crucial point is that the role of the quantum theory should be taken into consideration. According to quantum theory, vacuum is not "empty" in the simple sense of the word: it has rich physical import. The entire physical space is full of "virtual" particles. The role of these "virtual" particles can be verified through their physical effect. Under ordinary conditions "virtual" particles are being created and annihilated incessantly. There-

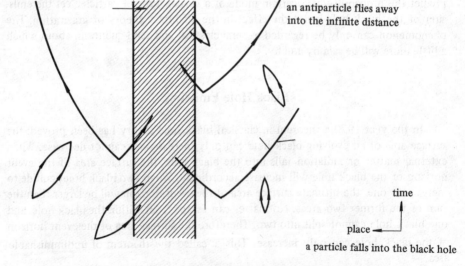

Fig. 12-1 In the vicinity of the black hole, one of the pair of particle and antiparticle may fall into the black hole, and the other particle, having lost its companion, can no longer be annihilated. This companionless particle may fly to the infinite distance. Then it seems as if the black hole were emitting particles or antiparticles

fore, vacuum cannot automatically produce particles or antiparticles. But when there is a gravitational field, especially a black hole, things will be different. Now if one of a pair of virtual electron and positron produced in vacuum falls into the black hole, it can never come out again, and the other one, having lost its companion, can never be annihilated again. This lone particle will either fall before long into the black hole, or fly away from its vicinity, the latter case being equivalent to an emission by the black hole (Fig.12—1). This is the vacuum emission evoked by the gravitational field of the black hole, the result of which will be the decrease of mass of the black hole.

The emission of the black hole is a kind of thermal emission: all of its spectra are of the black-body type. Therefore mo matter from what substance the black hole may trace its ancestry, we can only predict the probability of a certain particle in a certain mode, which conclusion we have mentioned already. But this theory is only concerned with the quantum fluctuation of vacuum. For the gravitational field of the black hole, it has also employed classical results. Therefore the theory as a whole is semi-classical, semi-quantum mechanical.

Superunification and Singularity

Though quantum gravitation theory is confronted with so many problems as mentioned above, the unification and quantization of gravitation has nevertheless made progress. The most promising way is to extend the general theory of relativity to the theory of supergravity. This theory of superunification or supersymmetry attempts to merge all inter actions into one. Moreover, it has a very fascinating feature: it has turn down the partition between the traditional "matter" and "interaction" in physics. In physics a half-integral-spin field is employed to represent "matter" particles, while the integral spin field is employed to represent "interaction" quanta, but in the theory of supersymmetry the half-intergral spin and the integral spin are unified. Moreover, in the theory of supergravity additional parametres are not introduced. Therefore, the theory is not only complete in unifying all the physical particles and interactions but also complete in that it admits no indefinite parameter whatever.

Yet do not think that once we have obtained a complete theory, we shall be able to know everything in principle. On the contrary, completeness means that something is definitely unknown rather than that everything is known. This is the singularity in theory.

In almost all kinds of theories singularities are to be met. The so-called singularity is none other than some "irrational" infinity. For instance, according to Newton's theory of gravitation, the gravitational potential ˙˙ goes to infinity in the infinite universe. Generally speaking, where infinity appears, there the theory is no longer applicable, but some more reasonable theory has to replace the old one. In truth, the development of the general theory of relativity has eliminated some infinities in Newton's theory of gravitation. This seems to suggest that with the finding of more and more correct and more and more complete theories, fewer and fewer singularities will stand in the way. But facts seem to be otherwise.

Though the general theory of relativity has done away with singularities in Newton's theory, it has also brought along with itself a series of singularities. The solution of the black hole contains a singularity; in cosmology there is also a singularity. The end of the gravitational collapse is a singularity, the beginning of the big bang is also a singularity.

For some time physicists believed that the singularities mentioned above were only a matter of mathematical form and could be shunned in reality. If we had not employed a geometrical structure of perfect symmetry, probably the singularity would not have turned up. Yet since the seventies a series of work has proved that in the general theory of relativity singularities are unavoidable: they have to be encountered in the course of evolution of the universe.

Some people even expect that the quantum theory of gravitation may probably eliminate these singularities. Now let us leave aside whether quantization can really work the miracle; the quantum effect of gravitation has already laid down the limit of 10^{28} eV which means the same as a singularity, thus making the relation of cause and effect "fade" into nought.

This recent development seems to demonstrate that with the deepening of the theoretical work, singularities still remain. Moreover, the higher perfection the theories have attained, the unavoidability of singularities or bounds of theory will become even more apparent.

Hence the bold conjecture: once we find the expected perfect unified theory of supergravity, we can trace back to the extremely early epoch—which may as well be called the singular epoch— of the universe during which the universe is younger than 10^{-43} sec. At that time there existed in the universe only one type of interaction, i.e. the "supergravitation". Only after the lapse of 10^{-43} sec, due to the "phase transition", did there appear in turn gravitation, strong interaction, weak interaction and electromagnetic interaction as we see them today. Therefore the very beginning of the big bang was a singularity, the real basic point of the universe where we live. If the universe is to be unified in this manner, is this singularity not the *ylem* which mankind has been searching for these two thousand years or more? As regards this alluring problem, it seems that we can only expect a new, but not the final, answer! This is because, if it is a singularity at all, then we cannot know everything about it. If we knew everything about it, it would cease to be a singularity.